LOGIC
and
REPRESENTATION

CSLI
Lecture Notes
No. 39

LOGIC
and
REPRESENTATION

Robert C. Moore

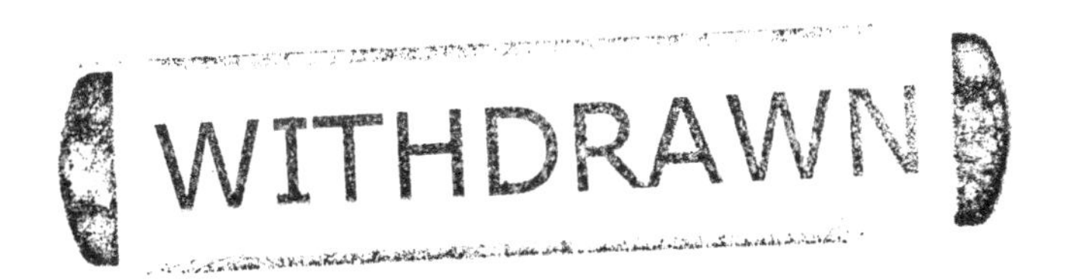

CENTER FOR THE STUDY OF
LANGUAGE AND INFORMATION
STANFORD, CALIFORNIA

CSLI was founded early in 1983 by researchers from Stanford University, SRI International, and Xerox PARC to further research and development of integrated theories of language, information, and computation. CSLI headquarters and the publication offices are located at the Stanford site.

CSLI/SRI International
333 Ravenswood Avenue
Menlo Park, CA 94025

CSLI/Stanford
Ventura Hall
Stanford, CA 94305

CSLI/Xerox PARC
3333 Coyote Hill Road
Palo Alto, CA 94304

Printed in the United States

99 98 97 96 95   5 4 3 2 1

Library of Congress Cataloging-in-Publication Data

Moore, Robert C., 1948–
Logic and Representation / Robert C. Moore.
p. cm. – (CSLI lecture notes ; no. 39)
Includes references and index.
ISBN 1–881526–16–X
ISBN 1–881526–15–1 (pbk.)
1. Language and logic. 2. Semantics (Philosophy) 3. Logic.
I.Title.
P39.M66 1995
160–dc20   94-40413
CIP

Contents

Acknowledgments

All the chapters of this book are edited versions of articles that have previously appeared elsewhere. Permission to use them here is gratefully acknowledged. Chapter 1 originally appeared under the title "The Role of Logic in Intelligent Systems," in *Intelligent Machinery: Theory and Practice*, ed. I. Benson, Cambridge, England: Cambridge University Press, 1986. Chapter 2 originally appeared in *The Behavioral and Brain Sciences*, Vol. 7, No. 4, 1984. Chapter 3 originally appeared in *Formal Theories of the Commonsense World*, ed. J. R. Hobbs and R. C. Moore, Norwood, New Jersey: Ablex Publishing Corporation, 1985. Chapter 4 originally appeared in *Processes, Beliefs, and Questions*, ed. S. Peters and E. Saarinen, Dordrecht, Holland: D. Reidel Publishing Company, 1982. Chapter 5 originally appeared in *Semantics and Contextual Expression*, ed. R. Bartch, J. van Benthem, and P. van Emde Boas, Dordrecht, Holland: Foris Publications, 1989. Chapter 6 originally appeared in *Artificial Intelligence*, Vol. 25, No. 1, 1985. Chapter 7 originally appeared in *Proceedings Non-Monotonic Reasoning Workshop*, New Paltz, New York, 1984. Chapter 8 originally appeared in *Artificial Intelligence*, Vol. 59, Nos. 1–2, 1993. Chapter 9 originally appeared in *EPIA 89, Proceedings 4th Portuguese Conference on Artificial Intelligence*, ed. J. P. Martins and E. M. Morgado, Berlin: Springer-Verlag, 1989. Chapter 10 originally appeared in *Proceedings 27th Annual Meeting of the Association for Computational Linguistics*, Vancouver, British Columbia, 1989.

These essays all reflect research carried out at SRI International, either in the Artificial Intelligence Center in Menlo Park, California, or the Computer Science Research Centre in Cambridge, England. I wish to thank the many SRI colleagues whose ideas, comments, and criticism over the years have influenced this work. I also owe a debt to numerous colleagues at other institutions, particularly the researchers from Stanford and Xerox PARC who came together with SRI to form CSLI in 1983.

I am grateful for a year spent as a fellow at the Center for Advanced Study in the Behavioral Sciences in 1979–80 as part of a special study group on Artificial Intelligence and Philosophy, supported by a grant from the Alfred P. Sloan Foundation. My interactions with the other Fellows in this group particularly influenced Chapters 2 and 5.

I also wish to thank my other research sponsors, who are cited individually in each chapter. Finally, I wish to thank Dikran Karagueuzian and his publications staff at CSLI for their efforts in pulling these texts together into a coherent whole, and for their patience during the long process.

Introduction

The essays collected in this volume represent work carried out over a period of more than ten years on a variety of problems in artificial intelligence, the philosophy of mind and language, and natural-language semantics, addressed from a perspective that takes as central the use of formal logic and the explicit representation of knowledge. The origins of the work could be traced even farther back than that, though, to the early 1970s when one of my goals as a graduate student was, in the *hubris* of youth, to write a book that would be the definitive refutation of Quine's *Word and Object* (1960). Over the intervening years I never managed to find the time to write the single extended essay that book was to have been, and more senior sages took on the task themselves in one way or another (with many of the resulting works being cited in these pages). In retrospect, however, I think that the point of view I wanted to put forth then largely comes through in these essays; so perhaps my early ambitions are at least partly realized in this work.

Two important convictions I have held on to since those early days are (1) that most of the higher forms of intelligent behavior require the explicit representation of knowledge and (2) that formal logic forms the cornerstone of knowledge representation. These essays show the development and evolution over the years of the application of those principles, but my basic views on these matters have changed relatively little. What has changed considerably more are the opposing points of view that are most prevalent. In the early 1970s, use of logic was somewhat in disrepute in artificial intelligence (AI), but the idea of explicit knowledge representation was largely unquestioned. In philosophy of mind and language, on the other hand, the idea of explicit representation of knowledge was just beginning to win its battle against the behaviorism of Quine and Skinner, powered by the intellectual energy generated by work in generative linguistics, AI, and cognitive psychology.

Today, in contrast, logic has made a comeback in AI to the point that, while it still has its critics, in the subfield of AI that self-consciously concerns itself with the study of knowledge representation, approaches based on logic have become the dominant paradigm. The idea of explicit knowledge representation itself, however, has come to be questioned by researchers working on neural networks (e.g., Rumelhart et al. 1987, McClelland et al. 1987) and reactive systems (e.g., Brooks 1991a, 1991b). In the philosophy of mind and language, the battle with behaviorism seems to be pretty much over (or perhaps I have just lost track of the argument).

In any case, I still find the basic arguments in favor of logic and representation as compelling as I did twenty years ago. Higher forms of human-like intelligence require explicit representation because of the recursive structure of the information that people are able to process. For any propositions P and Q that a person is able to contemplate, he or she is also able to contemplate their conjunction, "P and Q," their disjunction "P or Q," the conditional dependence of one upon the other "if P then Q," and so forth. While limitations of memory decrease our ability to reason with such propositions as their complexity increases, there is no reason to believe there is any architectural or structural upper bound on our ability to compose thoughts or concepts in this recursive fashion. To date, all the unquestioned successes of nonrepresentational models of intelligence have come in applications that do not require this kind of recursive structure, chiefly low-level pattern recognition and navigation tasks. No plausible models of tasks such as unbounded sentence comprehension or complex problem solving exist that do not rely on some form of explicit representation.

Recent achievements of nonrepresentational approaches, particularly in robot perception and navigation, are impressive, but claims that these approaches can be extended to higher-level forms of intelligence are unsupported by convincing arguments. To me, the following biological analogy seems quite suggestive: The perception and navigation abilities that are the most impressive achievements of nonrepresentational models are well within the capabilities of reptiles, which have no cerebral cortex. The higher cognitive abilities that seem to require representation exist in nature in their fullest form only in humans, who have by far the most developed cerebral cortex in the biological world. So, it would not surprise me if it turned out that in biological systems, explicit representations of the sort I am arguing for are constructed only in the cerebral cortex. This would suggest that there may be a very large role for nonrepresentational models of intelligence, but that they have definite limits as well.

Even if we accept that explicit representations are necessary for higher forms of intelligence, why must they be logical representations? That question is dealt with head-on in Chapter 1, but in brief, the argument is that only logical representations have the ability to represent certain forms of incomplete information, and that any representation scheme that has these abilities would *a fortiori* be a kind of logical representation.

Turning to the essays themselves, Part I consists of two chapters of a methodological character. Chapter 1 reviews a number of different roles for logic in AI. While the use of logic as a basis for knowledge representation is taken as central, elaborating the argument made above, the uses of logic as an analytical tool and as a programming language are also discussed. I might comment that it was only after this chapter was originally written that I gained much experience using PROLOG, the main programming language based on logic. Nevertheless, I find that my earlier analysis of logic programming holds up remarkably well, and I would change little if I were to re-write this chapter today. My current opinions are that the most useful feature of PROLOG is its powerful pattern-matching capability based on unification, that it is virtually impossible to write serious programs without going outside of the purely logical subset of the language, and that most of the other features of the language that derive from its origins in predicate logic get in the programmer's way more than they help.

Chapter 2 is a brief commentary that appeared as one of many accompanying a reprinting of Skinner's "Behaviorism at Fifty" (1984). Given the demise of behaviorism as a serious approach to understanding intelligence, it may be largely of historical interest, but it does lay out some of the basic counter arguments to classic behaviorist attacks on mentalistic psychology and mental representation.

Part II contains three chapters dealing with propositional attitudes, particularly knowledge and belief. Chapter 3 is a distillation of my doctoral dissertation, and presents a formal theory of knowledge and action. The goal of this work is to create a formal, general logic for expressing how the possibility of performing actions depends on knowledge and how carrying out actions affects knowledge. The fact that this logic is based on the technical constructs of possible-world semantics has misled some researchers to assume that I favored a theoretical analysis of propositional attitudes in terms of possible worlds. This has never been the case, however, and Chapters 4 and 5 present the actual development of my views on this subject.

Chapter 4 develops a semantics for belief reports (that is, statements like "John believes that P") based on a representational theory of belief. In the course of this development, a number of positive arguments for the representational theory of belief are presented that would fit quite comfortably among the methodological chapters in Part I. Later, I came to view the semantics proposed for propositional attitude reports in this chapter as too concrete, on the grounds that it would rule out the possibility of attributing propositional attitudes to other intelligent beings whose cognitive architecture was substantially different from our own. In its place, Chapter 5 presents a more abstract theory based on the notion of Russellian propositions. This chapter also provides a detailed comparison of this Russellian theory of attitude reports to the theory presented in the original version of situation semantics (Barwise and Perry 1983).

Part III presents three chapters concerning autoepistemic logic. This is a logic for modeling the beliefs of an agent who is able to introspect about his or her own beliefs. As such, autoepistemic logic is a kind of model of propositional attitudes, but it is distinguished from the formalisms discussed in Part II by being centrally concerned with how to model reasoning based on a *lack* of information. The ability to model this type of reasoning makes autoepistemic logic "nonmonotonic" in the sense of Minsky (1974). Chapter 6 presents the original work on autoepistemic logic as a rational reconstruction of McDermott and Doyle's nonmonotonic logic (1980, McDermott 1982). Chapter 7 presents an alternative, more formally tractable semantics for autoepistemic logic based on possible worlds, and Chapter 8 is a recently-written short retrospective surveying some of the subsequent work on autoepistemic logic and remaining problems.

Part IV consists of two essays on the topic of natural-language semantics. In taking a representational approach to semantics, we divide the problem into two parts; how to represent the meaning of natural-language expressions, and how to specify the mapping from language syntax into such a representation. Chapter 9 addresses the first issue from the standpoint of a set of problems concerning adverbial modifiers of action sentences. We compare two theories, one from Davidson (1967b) and one based on situation semantics (Perry 1983), concluding that aspects of both are needed for a full account of the phenomena. Chapter 10 addresses the problem of how to map between syntax and semantics, showing how a formalism based on the operation of unification can be a powerful tool for this purpose, and presenting a theoretical framework for compositionally interpreting the representations described by such a formalism.

Part I

Methodological Arguments

1

The Role of Logic in Artificial Intelligence

Formal logic has played an important part in artificial intelligence (AI) research for almost thirty years, but its role has always been controversial. This chapter surveys three possible applications of logic in AI: (1) as an analytical tool, (2) as a knowledge representation formalism and method of reasoning, and (3) as a programming language. The chapter examines each of these in turn, exploring both the problems and the prospects for the successful application of logic.

1.1 Logic as an Analytical Tool

Analysis of the content of knowledge representations is the application of logic in artificial intelligence (AI) that is, in a sense, conceptually prior to all others. It has become a truism to say that, for a system to be intelligent, it must have knowledge, and currently the only way we know of for giving a system knowledge is to embody it in some sort of structure—a *knowledge representation*. Now, whatever else a formalism may be, at least some of its expressions must have *truth-conditional semantics* if it is really to be a representation of knowledge. That is, there must be some sort of correspondence between an expression and the world, such that it makes sense to ask whether the world is the way the expression claims it to be. To have knowledge at all is to have knowledge[1] that the world is one way and not otherwise. If one's "knowledge" does not rule out any possibilities for how the world might be, then one really does not know anything at all. Moreover, whatever

Preparation of this chapter was made possible by a gift from the System Development Foundation as part of a coordinated research effort with the Center for the Study of Language and Information, Stanford University.

[1] Or at least a belief; most people in AI don't seem too concerned about truth in the actual world.

AI researchers may say, examination of their practice reveals that they do rely (at least informally) on being able to provide truth-conditional semantics for their formalisms. Whether we are dealing with conceptual dependencies, frames, semantic networks, or what have you, as soon as we say that a particular piece of structure represents the assertion (or belief, or knowledge) that John hit Mary, we have hold of something that is true if John did hit Mary and false if he didn't.

Mathematical logic (particularly model theory) is simply the branch of mathematics that deals with this sort of relationship between expressions and the world. If one is going to analyze the truth-conditional semantics of a representation formalism, then, *a fortiori*, one is going to be engaged in logic. As Newell puts it (1980, p. 17), "Just as talking of *programmerless* programming violates truth in packaging, so does talking of a *non-logical* analysis of knowledge."

While the use of logic as a tool for the analysis of meaning is perhaps the least controversial application of logic to AI, many proposed knowledge representations have failed to pass minimal standards of adequacy in this regard. (Woods (1975) and Hayes (1977) have both discussed this point at length.) For example, Kintsch (1974, p. 50) suggests representing "All men die" by (Die,Man) & (All,Man). How are we to evaluate such a proposal? Without a formal specification of how the meaning of this complex expression is derived from the meaning of its parts, all we can do is take the representation on faith. However, given some plausible assumptions, we can show that this expression cannot mean what Kintsch says it does.

The assumptions we need to make are that "&" means logical conjunction (i.e., "and"), and that related sentences receive analogous representations. In particular, we will assume that any expression of the form $(P \;\&\; Q)$ is true if and only if P is true and Q is true, and that "Some men dance" ought to be represented by (Dance,Man) & (Some,Man). If this were the case, however, "All men die" and "Some men dance" taken together would imply "All men dance." That, of course, does not follow, so we have shown that, if our assumptions are satisfied, the proposed representation cannot be correct. Perhaps Kintsch does not intend for "&" to be interpreted as "and," but then he owes us an explanation of what it *does* mean that is compatible with his other proposals.

Just to show that these model theoretic considerations do not simply lead to a requirement that we use standard logical notation, we can demonstrate that All(Man,Die) could be an adequate representation of "All men die." We simply let Man denote the set of all men, let Die denote the set of all things that die, and let All(X, Y) be true whenever the set denoted by X is a subset of the set denoted by Y. Then it will immediately follow that All(Men,Die) is true just in case all men die. Hence there is a systematic way of interpreting All(Men,Die) that is compatible with what it is claimed to mean.

The point of this exercise is that we want to be able to write computer programs whose behavior is a function of the meaning of the structures they manipulate. However, the behavior of a program can be directly influenced only by the form of those structures. Unless there is some systematic relationship between form and meaning, our goal cannot be realized.

1.2 Logic as a Knowledge Representation and Reasoning System

The Logic Controversy in AI

The second major application of logic to artificial intelligence is to use logic as a knowledge representation formalism in an intelligent computer system and to use logical deduction to draw inferences from the knowledge thus represented. Strictly speaking, there are two issues here. One could imagine using formal logic in a knowledge representation system, without using logical deduction to manipulate the representations, and one could even use logical deduction on representations that have little resemblance to standard formal logics; but the use of a logic as a representation and the use of logical deduction to draw inferences from the knowledge represented fit together in such a way that it makes most sense to consider them simultaneously.

This is a much more controversial application than merely using the tools of logic to analyze knowledge representation systems. Indeed, Newell (1980, p. 16) explicitly states that "the role of logic [is] as a tool for the analysis of knowledge, not for reasoning by intelligent agents." It is a commonly held opinion in the field that logic-based representations and logical deduction were tried many years ago and were found wanting. As Newell (1980, p. 17) expresses it, "The lessons of the sixties taught us something about the limitations of using logics for this role."

The lessons referred to by Newell were the conclusions widely drawn from early experiments in "resolution theorem-proving." In the mid 1960s, J. A. Robinson (1965) developed a relatively simple, logically complete method for proving theorems in first-order logic, based on the so-called resolution principle:[2]

[2]We will assume basic knowledge of first-order logic. For a clear introduction to first-order logic and resolution, see Nilsson (1980).

$$(P \lor Q), (\neg P \lor R) \models (Q \lor R)$$

That is, if we know that either P is true or Q is true and that either P is false or R is true, then we can infer that either Q is true or R is true.

Robinson's work brought about a rather dramatic shift in attitudes regarding the automation of logical inference. Previous efforts at automatic theorem-proving were generally thought of as exercises in expert problem solving, with the domain of application being logic, geometry, number theory, etc. The resolution method, however, seemed powerful enough to be used as a universal problem solver. Problems would be formalized as theorems to be proved in first-order logic in such a way that the solution could be extracted from the proof of the theorem.

The results of experiments directed towards this goal were disappointing. The difficulty was that, in general, the search space generated by the resolution method grows exponentially (or worse) with the number of formulas used to describe the problem and with the length of the proof, so that problems of even moderate complexity could not be solved in reasonable time. Several domain-independent heuristics were proposed to try to deal with this issue, but they proved too weak to produce satisfactory results. In the reaction that followed, not only was there was a turning away from attempts to use deduction to create general problem solvers, but there was also widespread condemnation of *any* use of logic in commonsense reasoning or problem-solving systems.

The Problem of Incomplete Knowledge

Despite the disappointments of the early experiments with resolution, there has been a recent revival of interest in the use of logic-based knowledge representation systems and deduction-based approaches to commonsense reasoning and problem solving. To a large degree this renewed interest seems to stem from the recognition of an important class of problems that resist solution by any other method.

The key issue is the extent to which a system has complete knowledge of the relevant aspects of the problem domain and the specific situation in which it is operating. To illustrate, suppose we have a knowledge base of personnel information for a company and we want to know whether any programmer earns more than the manager of data processing. If we have recorded in our knowledge base the job title and salary of every employee, we can simply find the salary of each programmer and compare it with the salary of the manager of data processing. This sort of "query evaluation" is essentially just an extended form of table lookup. No deductive reasoning is involved.

On the other hand, we might not have specific salary information in the knowledge base. Instead, we might have only general information such as "all programmers work in the data processing department, the manager of a department is the manager of all other employees of that department, and no employee earns more than his manager." From this information, we can deduce that no programmer earns more than the manager of data processing, although we have no information about the exact salary of any employee.

A representation formalism based on logic gives us the ability to represent information about a situation, even when we do not have a complete description of the situation. Deduction-based inference methods allow us to answer logically complex queries using a knowledge base containing such information, even when we cannot "evaluate" a query directly. On the other hand, AI inference systems that are not based on automatic-deduction techniques either do not permit logically complex queries to be asked, or they answer such queries by methods that depend on the possesion of complete information.

First-order logic can represent incomplete information about a situation by

Saying that something has a certain property without saying which thing has that property: $\exists x P(x)$

Saying that everything in a certain class has a certain property without saying what everything in that class is: $\forall x(P(x) \supset Q(x))$

Saying that at least one of two statements is true without saying which statement is true: $(P \vee Q)$

Explicitly saying that a statement is false, as distinguished from not saying that it is true: $\neg P$

These capabilities would seem to be necessary for handling the kinds of incomplete information that people can understand, and thus they would be required for a system to exhibit what we would regard as general intelligence. Any representation formalism that has these capabilities will be, at the very least, an extension of classical first-order logic, and any inference system that can deal adequately with these kinds of generalizations will have to have at least the capabilities of an automatic-deduction system.

The Control Problem in Deduction

If the negative conclusions that were widely drawn from the early experiments in automatic theorem-proving were fully justified, then we would have a virtual proof of the impossibility of creating intelligent systems based on the knowledge representation approach, since many types of incomplete knowledge that people are capable of dealing with seem to demand the use of logical representation and deductive inference. A careful analysis, however, suggests that the failure of the early attempts to do commonsense reasoning and problem solving by theorem-proving had more specific causes that can be attacked without discarding logic itself.

The point of view we shall adopt here is that there is nothing wrong with using logic or deduction per se, but that a system must have some way of knowing, out of the many possible inferences it could draw, which ones it *should* draw. A very simple, but nonetheless important, instance of this arises in deciding how to use assertions of the form $P \supset Q$ ("P implies Q"). Intuitively, such a statement has at least two possible uses in reasoning. Obviously, one way of using $P \supset Q$ is to infer Q, whenever we have inferred P. But $P \supset Q$ can also be used, even if we have not yet inferred P, to suggest a way to infer Q, if that is what we are trying to do. These two ways of using an implication are referred to as *forward chaining* ("If P is asserted, also assert Q") and *backward chaining* ("To infer Q, try to infer P"), respectively. We can think of the deductive process as a bidirectional search, partly working forward from what we already know, partly working backward from what we would like to infer, and converging somewhere in the middle.

Unrestricted use of the resolution method turns out to be equivalent to using every implication both ways, leading to highly redundant searches. Domain-independent refinements of resolution avoid some of this redundancy, but usually impose uniform strategies that may be inappropriate in particular cases. For example, often the strategy is to use all assertions only in a backward-chaining manner, on the grounds that this will at least guarantee that all the inferences drawn are relevant to the problem at hand.

The difficulty with this approach is that whether it is more efficient to use an assertion for forward chaining or for backward chaining can depend on the specific form of the assertion, or the set of assertions in which it is embedded. Consider, for instance, the following schema:

$$\forall x(P(F(x)) \supset P(x))$$

Instances of this schema include such things as:

$$\forall x(x+1 < y \supset x < y)$$
$$\forall x(\mathsf{Jewish}(\mathsf{Mother}(x)) \supset \mathsf{Jewish}(x))$$

That is, a number x is less than a number y if $x+1$ is less than y; and a person is Jewish if his or her mother is Jewish.[3]

[3] I am indebted to Richard Waldinger for suggesting the latter example.

Suppose we were to try to use an assertion of the form $\forall x(P(F(x)) \supset P(x))$ for backward chaining, as most "uniform" proof procedures would. In effect, we would have the rule, "To infer $P(x)$, try to infer $P(F(x))$." If, for instance, we were trying to infer $P(A)$, this rule would cause us to try to infer $P(F(A))$. This expression, however, is also of the form $P(x)$, so the process would be repeated, resulting in an infinite descending chain of formulas to be inferred:

$P(A)$
$P(F(A))$
$P(F(F(A)))$
$P(F(F(F(A))))$, etc.

If, on the other hand, we use the rule for forward chaining, the number of applications is limited by the complexity of the assertion that originally triggers the inference. Asserting a formula of the form $P(F(x))$ would result in the corresponding instance of $P(x)$ being inferred, but each step reduces the complexity of the formula produced, so the process terminates:

$P(F(F(A)))$
$P(F(A))$
$P(A)$

It turns out, then, that the efficent use of a particular assertion often depends on exactly what that assertion is, as well as on the context of other assertions in which it is embedded. Kowalski (1979) and Moore (1980b) illustrate this point with examples involving not only the distinction between forward chaining and backward chaining, but other control decisions as well.

In some cases, control of the deductive process is affected by the details of how a concept is axiomatized, in ways that go beyond "local" choices such as that between forward and backward chaining. Sometimes logically equivalent formalizations can have radically different behavior when used with standard deduction techniques. For example, in the blocks world that has been used as a testbed for so much AI research, it is common to define the relation "A is Above B" in terms of the primitive relation "A is (directly) On B," with Above being the transitive closure of On. This can be done formally in at least three ways:[4]

$$\forall x, y(\mathsf{Above}(x, y) \equiv (\mathsf{On}(x, y) \vee \exists z(\mathsf{On}(x, z) \wedge \mathsf{Above}(z, y))))$$

[4] These formalizations are not quite equivalent, as they allow for different possible interpretations of Above, if infinitely many objects are involved. They are equivalent, however, if only a finite set of objects is being considered.

$$\forall x, y(\mathsf{Above}(x, y) \equiv (\mathsf{On}(x, y) \vee \exists z(\mathsf{Above}(x, z) \wedge \mathsf{On}(z, y))))$$
$$\forall x, y(\mathsf{Above}(x, y) \equiv (\mathsf{On}(x, y) \vee \exists z(\mathsf{Above}(x, z) \wedge \mathsf{Above}(z, y))))$$

Each of these axioms will produce different behavior in a standard deduction system, no matter how we make such local control decisions as whether to use forward or backward chaining. The first axiom defines Above in terms of On, in effect, by iterating upward from the lower object, and would therefore be useful for enumerating all the objects that are above a given object. The second axiom iterates downward from the upper object, and could be used for enumerating all the objects that a given object is above. The third axiom, though, is essentially a "middle out" definition, and is hard to control for any specific use.

The early systems for problem solving by theorem-proving were often inefficient because axioms were chosen for their simplicity and brevity, without regard to their computational properties—a problem that also arises in conventional programming. To take a well-known example, the simplest procedure for computing the nth Fibonacci number is a doubly recursive algorithm whose execution time is proportional to 2^n, while a slightly more complicated, less intuitively defined, singly recursive procedure can compute the same function time proportional to n.

Prospects for Logic-Based Reasoning Systems

The fact that the issues discussed in this section were not taken into account in the early experiments in problem solving by theorem-proving suggests that not too much weight should be given to the negative results that were obtained. As yet, however, there is not enough experience with providing explicit control information and manipulating the form of axioms for computational efficiency to tell whether large bodies of commonsense knowledge can be dealt with effectively through deductive techniques. If the answer turns out to be "no," then some radically new approach will be required for dealing with incomplete knowledge.

1.3 Logic as a Programming Language

Computation and Deduction

The parallels between the manipulation of axiom systems for efficient deduction and the design of efficient computer programs were recognized in the early 1970s by a number of people, notably Hayes (1973), Kowalski (1974), and Colmerauer (1978). It was discovered, moreover, that there are ways to formalize many functions and relations so that the application of standard deduction methods will have the effect of executing them as efficient computer programs. These observations have led to the development of the field of logic programming and the creation of new computer languages such as PROLOG (Warren, Pereira, and Pereira 1977).

As an illustration of the basic idea of logic programming, consider the "append" function, which appends one list to the end of another. This function can be implemented in LISP as follows:

```
(append  a b) =
(cond((null a b)
        (t (cons (car a) (append (cdr a) b))))
```

What this function definition says is that the result of appending B to the end of A is B if A is the empty list, otherwise it is a list whose first element is the first element of A and whose remainder is the result of appending B to the remainder of A.

We can easily write a set of axioms in first-order logic that explicitly say what we just said in English. If we treat Append as a three-place relation (with $\mathsf{Append}(A, B, C)$ meaning that C is the result of appending B to the end of A) the axioms might look as follows[5] :

$$\forall x(\mathsf{Append}(\mathsf{Nil}, x, x)$$
$$\forall x, y, z(\mathsf{Append}(x, y, z) \supset \forall w(\mathsf{Append}(\mathsf{Cons}(w, x), y, \mathsf{Cons}(w, z))))$$

The key observation is that, when these axioms are used via backward chaining to infer $\mathsf{Append}(A, B, x)$, where A and B are arbitrary lists and x is a variable, the resulting deduction process not only terminates with the variable x bound to the result of appending B to the end of A, it exactly mirrors the execution of the corresponding LISP program. This suggests that in many cases, by controlling the use of axioms correctly, deductive methods can be used to simulate ordinary computation with no loss of efficiency. The new view of the relationship between deduction and computation that emerged from these observations was, as Hayes (1973) put it, "Computation is controlled deduction."

The ideas of logic programming have produced a very exciting and fruitful new area of research. However, as with all good new ideas, there has been a degree of "over-selling" of logic programming and, particularly, of the PROLOG language. So, if the following sections focus more on the limitations of logic programming than on its strengths,

[5] To see the equivalence between the LISP program and these axioms, note that $\mathsf{Cons}(w, x)$ corresponds to A, so that w corresponds to (car A) and x corresponds to (cdr A).

they should be viewed as an effort to counterbalance some of the overstated claims made elsewhere.

Logic Programming and PROLOG

To date, the main application of the idea of logic programming has been the development of the programming language PROLOG. Because it has roots both in programming methodology and in automatic theorem-proving, there is a widespread ambivalence about how PROLOG should be viewed. Sometimes it is seen as "just a programming language," although with some very interesting and useful features, and other times it is viewed as an "inference engine," which can be used directly as the basis of a reasoning system. On occasion these two ways of looking at PROLOG are simply confused, as when the (false) claim is made that to program in PROLOG one has simply to state the facts of the problem one is trying to solve and the PROLOG system will take care of everything else. This confusion is also evident in the terminology associated with the Japanese fifth generation computer project, in which the basic measure of machine speed is said to be "logical inferences per second." We will try to separate these two ways of looking at PROLOG, evaluating it first as a programming language and then as an inference system.

To evaluate PROLOG as a programming language, we will compare it with LISP, the programming language most widely used in AI.[6] PROLOG incorporates a number of features not found in LISP:

Failure-driven backtracking
Procedure invocation by pattern matching (unification)
Pattern matching as a substitute for selector functions
Procedures with multiple outputs
Returning and passing partial results via structures containing logical variables

These features and others make PROLOG an extremely powerful language for certain applications. For example, its incorporation of backtracking, pattern matching, and logical variables make it ideal for the implementation of depth-first parsers for language processing.[7] It is probably impossible to do this as efficiently in LISP as in PROLOG.

---

[6] The fact that the idea of logic programming grew out of AI work on automated inference, of course, gives AI no special status as a domain of application for logic programming. But because it was developed by people working in AI, and because it provides good facilities for symbol manipulation, most PROLOG applications have been within AI.

[7] This is in fact the application for which it was invented.

Moreover, having pattern matching as the standard way of passing information between procedures and decomposing complex structures makes many programs much simpler to write and understand in PROLOG than in LISP. On the other hand, PROLOG lacks general purpose operators for changing data structures. In applications where such facilities are needed, such as maintaining a highly interconnected network structure, PROLOG can be awkward to use. For this type of application, using LISP is much more straightforward.

To better understand the advantages and disadvantages of PROLOG relative to LISP, it is helpful to consider that PROLOG and LISP both contain a purely declarative subset, in which every expression affects the course of a computation only by its value, not by "side effects." For example, evaluating $(2 + 3)$ would normally not change the computational state of the system, while evaluating $(X \leftarrow 3)$ would change the value of X. In comparing their "pure" subsets, one finds that PROLOG is strictly more general than LISP. These subsets can both be thought of as logic programming languages, but the logic of pure LISP is restricted to recursive function definitions, while that of PROLOG permits definitions of arbitrary relations. This is what gives rise to the use of backtracking control structure, multiple return values, and logical variables. Pure PROLOG, then, can be thought of as a conceptual extension of pure LISP.

The creators of LISP, however, recognized that "although this language [pure LISP] is universal in terms of computable functions of symbolic expressions, it is not convenient as a programming system without additional tools to increase its power," (McCarthy et al 1962, p. 41). What was added to LISP was a set of operations for directly manipulating the pointer structures that represent the abstract symbolic expressions forming the semantic domain of pure LISP. LISP thus operates at two distinct levels of abstraction; simple things can be done quite elegantly at the level of recursive functions of symbolic expressions, while more complex tasks can be dealt with at the level of operations on pointer structures. Both levels, though, are conceptually coherent and, in a sense, complete.

PROLOG also has extensions to its purely logical core that most users agree are essential to its use as practical programming language. These extensions, however, do not have the kind of uniform conceptual basis that the structure manipulation features of LISP do. Such features as the "cut" operation for terminating backtracking, "assert" and "retract" for altering the PROLOG database, and predicates that test whether variables are free or bound are all powerful and useful devices, but they do not share any common semantic domain of operation. There is nothing categorically objectionable about any of these features in isolation, but they do not fit together in a coherent way. The result is that, while PROLOG provides a very powerful set of tools, the effective use of those tools depends to a greater extent than with many other languages on the ingenuity of the programmer and his acquaintance with the lore of the user community.[8]

This suggests that if PROLOG is really to replace LISP as the language of choice for AI systems, it should be given a more powerful and more conceptually coherent set of nonlogical extensions to the basic logic-programming paradigm, analogous to LISP's nonlogical extensions to the recursive-function paradigm. This suggestion would no doubt be resisted by purists who see the present nonlogical features of PROLOG as already departing too far from the semantic elegance of a system where the correctness of a program can be judged simply by whether all of its statements are *true*; but that is an idealized vision whose practical realization is doubtful.[9]

PROLOG as an Inference System

Whatever its merits purely as a programming language, much of the current enthusiasm for PROLOG undoubtedly stems from the impression that, because a PROLOG interpreter can be viewed as an automatic theorem-prover, PROLOG itself can be used as the reasoning module of an intelligent system. This is true to an extent, but only to a limited extent. The major limitation is that all practical logic programming systems to date, including PROLOG, are based, not on full first-order logic, but on the *Horn-clause* subset of first-order logic.

The easiest way to view Horn-clause logic is to say that axioms must be either atomic formulas such as $\mathsf{On}(A, B)$ or implications whose consequent is an atomic formula and whose antecedent is either an atomic formula or a conjunction of atomic formulas:

---

[8] To be fair, this last statement is true of LISP as well, especially with regard to recent extensions, such as "flavors." But it seems that with PROLOG one is forced into this domain of semantic uncertainty sooner than with LISP.

[9] One can make a plausible argument that the advent of massively parallel computer architectures will change this situation. For the type of problem that would normally be solved by an algorithm that changes data structures, using an imperative language typically requires fewer computation steps than using a declarative language but creates more timing dependencies. Thus parallel architectures and declarative languages are well matched, because the architecture provides the greater computational resources required by the language, and the language provides the lack of timing dependencies required to take advantage of the architecture. It remains to be seen, however, for how wide a class of problems the speedups due to parallelism outweigh the additional computation steps required.

$$(\mathsf{On}(x, y) \wedge \mathsf{Above}(y, z)) \supset \mathsf{Above}(x, z)$$

Furthermore, the only queries that can be posed are those that can be expressed as a disjunction of conjunctions of atomic formulas:

$$(\mathsf{On}(A, B) \wedge \mathsf{On}(B, C)) \vee (\mathsf{On}(C, B) \wedge \mathsf{On}(B, A))$$

These limitations mean that no negative formulas—for example, $\neg\mathsf{On}(A, B)$—can ever be asserted or inferred, and no disjunction can be inferred unless one of the disjuncts can be inferred. Thus, Horn-clause logic gives up two of the main features of first-order logic that permit reasoning with incomplete knowledge: being able to say or infer that one of two statements is true without knowing which is true, and being able to distinguish between knowing that a statement is false and not knowing that it is true.

The question of quantification is more complicated. Horn-clause logic does not permit quantifiers per se, but it does allow formulas to contain function symbols and free variables, and there is a result (Skolem's theorem) to the effect that with these devices, any quantified formula can be replaced by one without quantifiers. However, this quantifier-elimination theorem does not apply to most logic programming systems, because of the way they implement unification (pattern matching).

According to the usual mathematical definition of unification, a variable cannot be unified with any expression in which it is a proper subexpression. That is, x will not unify with $F(G(x))$, because there is no fully instantiated value for x that will make these two expressions identical. The test for this condition is usually called "the occur check." The occur check is computationally expensive, though, so most logic programming systems omit it for the sake of efficiency. There is a mathematically rigorous foundation for unification operation without the occur check, based on infinite trees, but this version of unification is *not* compatible with the quantifier-elimination techniques usually used in automatic theorem-proving. In particular, without the occur check, a logic programming system cannot properly distinguish between formulas that differ only in quantifier scope, such as, $\forall x(\exists y(P(x, y)))$ and $\exists y(\forall x(P(x, y)))$. That is, the system cannot distinguish between the statement that every person has a mother, and the statement that every person has the *same* mother.

These restrictions are so severe that PROLOG is almost never used as a reasoning system without using the extra-logical features of the language to augment its expressive power. In particular, the usual practice is to define negation in the system, using the "cut" operation, so that $\neg P$ can be inferred by having an attempt to infer P terminate in failure. Making this extension permits the implementation of nontrivial reasoning systems in PROLOG in a very direct way,[10] but it amounts to making "the closed-world assumption": any statement that cannot be inferred to be true is assumed to be false. To adopt this principle, though, is to give up entirely on trying to reason with incomplete knowledge, which is the main advantage that logic-based systems have over their rivals.

To see what one gives up in making the closed-world assumption, consider the following problem, adapted from Moore (1980b, p. 28). Three blocks, A, B, and C, are arranged as shown:

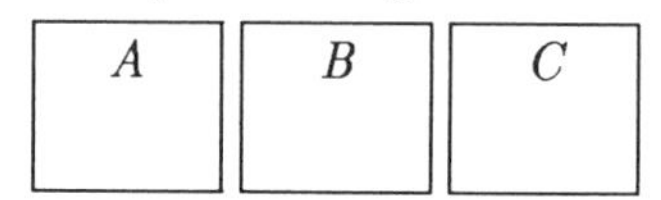

A is green, C is blue, and the color of B is unstated. In this arrangement of blocks, is there a green block next to a block that is not green? It should be clear with no more than a moment's reflection that the answer is "yes." If B is green, it is a green block next to the nongreen block C; if B is not green then A is a green block next to the nongreen block B.

To solve this problem, a reasoning system must be able to withold judgment on whether block B is green; it must know that either B is green or B is not green without knowing which; and it must use this fact to infer that *some* blocks stand in a certain relation to each other, without being able to infer which blocks these are. None of this is possible in a system that makes the closed-world assumption.

This is not to say that using PROLOG as a reasoning system with the closed-world assumption is always a bad thing to do. For applications where the closed-world assumption is justified, using PROLOG in this way can be extremely efficient—possibly more efficient than anything that can be programmed in LISP (for much the same reasons that top-down parsing is so efficient in PROLOG). But not all situations justify the closed-world assumption, and where it is not justified, the fact that PROLOG can be viewed as a theorem-prover is irrelevant. The usefulness of PROLOG in such a case will depend only on its utility as a programming language for implementing other inference systems.

1.4 Conclusions

In this chapter we have reviewed three possible applications of formal logic in artificial intelligence: as a tool for analyzing knowledge-

[10] Ironically, it is necessary to go outside the purely logical subset of PROLOG to do this!

representation formalisms, as a source of representation formalisms and reasoning methods, and as a programming language. As an analytical tool, the mathematical framework developed in the study of formal logics is simply the only tool we have for analyzing anything *as* a representation. There is little more to say, other than to note all the efforts to devise representation formalisms that have come to grief for lack of adequate logical analysis.

The other two applications are more controversial. A large segment of the AI community believes that any representation or deduction system based on standard logic will necessarily be too inefficient to be of any practical value. We have argued that such negative conclusions are based on experiments in which there was insufficient control of the deductive process, and we have presented a number of cases in which better control would lead to more efficient processing. Moreover, we have argued that when an application involves incomplete knowledge of the problem, only systems based on logic seem adequate to the task.

The use of logic as a basis for programming languages is the most recent application of logic within AI. We had two major points to make in this area. First, current logic programming languages (i.e., PROLOG) need to be more developed in their *non*logical features before they can really replace LISP as the primary language for developing intelligent systems. Second, as they currently exist, logic programming languages are suitable for direct use as inference systems only in a very restricted class of applications.

After thirty years, where does the use of logic in AI now stand? In all fairness, would one have to say that its promise has yet to be proven—but, of course, that is true for most of the field of AI. It may be that, if the promise of logic is to be fulfilled, it will have to come in a remerging of two of the main themes explored in this chapter: automatic deduction and logic programming. Logic programming grew out of the realization that, if automated reasoning systems are to perform efficiently, the information they are given must be carefully structured in much the same way that efficient computer programs are structured. But, instead of using that insight to produce more efficient reasoning systems, the developers of logic programming applied their ideas to more conventional programming problems. Perhaps the time is now right to take what has been learned about the efficient use of logic in logic programming, and apply it to the more general use of logic in automated reasoning. This just might produce the kind of basic technology for reasoning systems on which the development of the entire field depends.

2

A Cognitivist Reply to Behaviorism

In "Behaviorism at Fifty," B. F. Skinner (1984) attacks the idea of mentalistic psychology in general, and mental representation in particular. There are two major themes running through Skinner's various objections. He argues, first, that mentalistic notions have no explanatory value ("The objection is not that these things are mental, but that they offer no real explanation..."), and second, that since the correct explanation of behavior is in terms of stimuli and responses, mentalistic accounts of behavior must be either false or translatable into behavioristic terms ("...behavior which seemed to be the product of mental activity could be explained in other ways."). What I hope to show is that a "cognitivist" perspective offers a way of constructing mentalistic psychological theories that circumvent both kinds of objection.

The first theme appears twice in infinite-regress arguments. Skinner ridicules psychological theories that seem to appeal to homunculi, on the grounds that explaining the behavior of one homunculus would require a second homunclus, and so on. Later he employs the same rationale to criticize theories of perception based on internal representation: If seeing consists of constructing an internal representation of the thing seen, the internal representation would then apparently require an inner eye to look at it, etc. Skinner's concern for explanatory value is also evident in his view of mental states as mere "way stations" in unfinished causal accounts of behavior. If an act is said to have been caused by a certain mental state, without any account as to how that state itself was caused, there seems to be little to constrain

Preparation of this chapter was made possible by a gift from the System Development Foundation as part of a coordinated research effort with the Center for the Study of Language and Information, Stanford University.

what states we invoke to explain behavior. The limiting case would be to "explain" every action an agent performs by simply postulating a primitive desire to perform that action.

Skinner's concerns about explanatory value should not be taken lightly, and they seem to me to pose serious problems for older-style mentalistic psychological theories. Often these theories appear to allow no direct evidence for the existence of many kinds of mental states and events. According to such theories, "poking around the brain" will not help, because mental entities are not physical; moreover, asking the subject for introspective reports may not help either, because mental entities can be unconscious. But a second consequence of the view that mental entities are nonphysical is that we have no a priori idea as to what the constraints on their causal powers might be. We are thus left in a situation in which we could, at least in principle, postulate any mental states and events we like, adjusting our assumptions regarding their effects on behavior to fit any possible evidence.

How does cognitivism avoid Skinner's charges in this area? I take it that what distinguishes cognitivism from other mentalistic approaches to psychology is the premise that mental states can be identified with *computational* states. This has two consequences for the problem at hand. First, computational states must in some way be embodied in physical states. This means that if behavioral evidence alone were not sufficient to determine what mental state an organism was in, neurological evidence could be brought to bear to decide the question. Second, and of much more immediate practical consequence, is the fact that there is a very well-developed mathematical theory of the abilities and limits of computational systems. Hence, once we identify mental states with computational states, we are not free to endow them with arbitrary causal powers.

When a computational account of mental states and events is given, Skinner's infinite-regress arguments lose their force. While it *is* a characteristic of computational theories of mind to explain the behavior of the whole organism in terms of interactions among systems that may appear to be "homunculi," a computational account, as Dennett (1978, p. 123–124) has pointed out, requires each of these homunculi to be less intelligent than the whole they comprise. Thus, while there is indeed a regress, it is not an infinite one, because eventually we get down to a level of homunculi so stupid that they can be clearly seen to be "mere machines." Similar comments apply to Skinner's worries about explaining perception in terms of mental representation. Although he is quite correct in maintaining the pointlessness of supposing that the brain contains an isomorphic copy of the image on the retina, computational theories of vision simply do not work that way. Although they make use of internal representations, these express an *interpretation* of the image, not a copy. While a retinal image might be thought of as a two-dimensional array of light intensities, the postulated representations take as primitives such notions as "convex edge," "concave edge," and "occluding edge." These representations are then manipulated computationally in ways that make sense given their interpretations. Waltz (1975) gives a very clear (albeit already outdated) exposition of this approach.

Skinner's notion of unfinished causal account is not necessarily answered simply by adopting a computational perspective, but conscientious cognitive theorists do address the problems raised by the tendency to attribute precisely those structures that are needed to account for observed behavior. Some deal with it as Skinner suggests, by investigating the causation of mental states (e.g. studying language acquisition), but the more frequent strategy is to show how a single computational mechanism (or the interaction of a few mechanisms) accounts for a broad range of behavior. If, for example, we can show that a relatively small set of linguistic rules can account for a much larger (perhaps infinite) set of natural-language sentence patterns, then it is certainly not vacuous, or without explanatory value, to claim that those linguistic rules in some sense characterize the mental state of a competent language user.

Whether or not Skinner would acknowledge that the cognitivist framework has the potential to produce mentalistic theories with genuine explanatory value, I suspect he would argue that, because of the other major theme of his paper, any such conclusion is really beside the point. In his view, mentalistic terminology is at best a rather complicated and misleading way of talking about behavior and behavioral dispositions. Skinner's picture seems to be that mental states, rather than being real entities that mediate between stimulus and response, are merely *summaries* of stimulus-response relationships. Thus, hunger, rather than being what causes us to eat when presented with food, would be regarded as the disposition to eat when presented with food. (This interpretation of mental states obviously reinforces Skinner's opinion that mental explanations of behavior are vacuous; attributing eating to a disposition to eat explains nothing.)

The response to this point of view is that, even if we could get a complete description of an organism's "mental state" in terms of behavioral dispositions, that fact would not vitiate attempts to give a causal account of those dispositions in a way that might make reference to mental states more realistically construed. A computer analogy is helpful here. Complex computer systems often have "users' manuals" that are intended, in effect, to be complete accounts of the systems' behavioral dispositions. That is, they undertake to describe for any input (stimulus) what the output (response) of the system would be. But no one would suppose that to know the content of the user's manual is to know everything about a system; we might not know anything at all about how the system achieves the behavior described in the manual. Skinner's response might be that, if we want to know how the behavioral dispositions of an organism are produced, we have to look to neurobiology—but this would miss the point of one of the most important substantive claims of cognitivism. Just as in a complex computer system there are levels of abstraction above the level of electronic components (the analogue, one supposes, of neurons) that comprise coherent domains of discourse in which causal explanations of behavior can be couched ("The system computes square roots by Newton's method."), so too in human psychology there seem to be similar levels of abstraction—including levels that involve structures corresponding roughly to such pretheoretical mentalistic concepts as belief, desire, and intention.

Finally, it may very well be *impossible* to describe the behavioral dispositions of organisms as complex as human beings without reference to internal states. Skinner seems to assume uncritically that, if the sole objective of psychology is to describe the stimulus-response behavior of organisms, one can always do so without reference to internal states. But this is mathematically impossible for many of the formal models one might want to use to describe human behavior. In particular, given some of the behavioral repertoires that human beings are capable of acquiring (e.g., proving theorems in mathematics, understanding the well-formed expressions of a natural language), it seems likely that no formal model significantly less powerful than a general-purpose computer (Turing machine) could account for the richness of human behavior. In a very strong sense, however, it is generally impossible to characterize the behavior of a Turing machine without referring to its internal states. Now, the behaviorists may be fortunate, and it may turn out that the behavioral dispositions of humans are indeed describable without reference to internal states, but Skinner appears not even to realize that this is a problem.

To summarize: (1) Skinner's arguments against the explanatory value of mentalistic psychology do not apply to properly constructed cognitivist theories; (2) the existence of a complete behavioristic psychology would neither supplant nor render superfluous a causal cognitivist account of psychology; (3) the regularities of human behavior that Skinner's approach to psychology attempts to describe may not even be expressible without reference to internal states.

Part II

Propositional Attitudes

3

A Formal Theory of Knowledge and Action

3.1 The Interplay of Knowledge and Action

Most work on planning and problem solving within the field of artificial intelligence assumes that the agent has complete knowledge of all relevant aspects of the problem domain and problem situation. In the real world, however, planning and acting must frequently be performed without complete knowledge. This imposes two additional burdens on an intelligent agent trying to act effectively. First, when the agent entertains a plan for achieving some goal, he must consider not only whether the physical prerequisites of the plan have been satisfied, but also whether he has all the information necessary to carry out the plan. Second, he must be able to reason about what he can do to obtain necessary information that he lacks. In this chapter, we present a theory of action in which these problems are taken into account, showing how to formalize both the knowledge prerequisites of action and the effects of action on knowledge.

Planning sequences of actions and reasoning about their effects is one of the most thoroughly studied areas within artificial intelligence (AI). Relatively little attention has been paid, however, to the important role that an agent's knowledge plays in planning and acting to achieve a goal. Virtually all AI planning systems are designed to op-

The research reported herein was supported in part by the Air Force Office of Scientific Research under Contract No. F49620-82-K-0031. The views and conclusions expressed in this document are those of the author and should not be interpreted as necessarily representing the official policies or endorsements, either expressed or implied, of the Air Force Office of Scientific Research of the U.S. Government. This research was also made possible in part by a gift from the System Development Foundation as part of a coordinated research effort with the Center for the Study of Language and Information, Stanford University.

erate with complete knowledge of all relevant aspects of the problem domain and problem situation. Often any statement that cannot be inferred to be true is assumed to be false. In the real world, however, planning and acting must frequently be performed without complete knowledge of the situation.

This imposes two additional burdens on an intelligent agent trying to act effectively. First, when the agent entertains a plan for achieving some goal, he must consider not only whether the physical prerequisites of the plan have been satisfied, but also whether he has all the information necessary to carry out the plan. Second, he must be able to reason about what he can do to obtain necessary information that he lacks. AI planning systems are usually based on the assumption that, if there is an action an agent is physically able to perform, and carrying out that action would result in the achievement of a goal P, then the agent can achieve P. With goals such as opening a safe, however, there are actions that any human agent of normal abilities is physically capable of performing that would result in achievement of the goal (in this case, dialing the combination of the safe), but it would be highly misleading to claim that an agent could open a safe simply by dialing the combination unless he actually *knew* that combination. On the other hand, if the agent had a piece of paper on which the combination of the safe was written, he could open the safe by reading what was on the piece of paper and then dialing the combination, even if he did not know it previously.

In this chapter, we will describe a formal theory of knowledge and action that is based on a general understanding of the relationship between the two.[1] The question of generality is somewhat problematical, since different actions obviously have different prerequisites and results that involve knowledge. What we will try to do is to set up a formalism in which very general conclusions can be drawn, once a certain minimum of information has been provided concerning the relation between specific actions and the knowledge of agents.

To see what this amounts to, consider the notion of a test. The essence of a test is that it is an action with a directly observable result that depends conditionally on an unobservable precondition. In the use of litmus paper to test the pH of a solution, the observable result is whether the paper has turned red or blue, and the unobservable precondition is whether the solution is acid or alkaline. What makes

[1]This chapter presents the analysis of knowledge and action, and the representation of that analysis in first-order logic, that were developed in the author's doctoral thesis (Moore 1980a). The material in Section 3.3, however, has been substantially revised.

such a test useful for acquiring knowledge is that the agent can infer whether the solution is acid or alkaline on the basis of his knowledge of the behavior of litmus paper and the observed color of the paper. When one is performing a test, it is this inferred knowledge, rather than what is directly observed, that is of primary interest.

If we tried to formalize the results of such a test by making simple assertions about what the agent knows subsequent to the action, we would have to include the result that the agent knows whether the solution is acid or alkaline as a separate assertion from the result that he knows the color of the paper. If we did this, however, we would completely miss the point that knowledge of the pH of the solution is inferred from other knowledge, rather than being a direct observation. In effect, we would be *stipulating* what actions can be used as tests, rather than creating a formalism within which we can *infer* what actions can be used as tests.

If we want a formal theory of how an agent's state of knowledge is changed by his performing a test, we have to represent and be able to draw inferences from the agent's having several independent pieces of information. Obviously, we have to represent that, after the test is performed, the agent knows the observable result. Furthermore, we have to represent the fact that he knows that the test has been performed. If he just walks into the room and sees the litmus paper on the table, he will know what color it is, but, unless he knows its recent history, he will not have gained any knowledge about the acidity of the solution. We also need to represent the fact that the agent understands how the test works; that is, he knows how the observable result of the action depends on the unobservable precondition. Even if he sees the litmus paper put into the solution and then sees the paper change color, he still will not know whether the solution is acid or alkaline unless he knows how the color of the paper is related to the acidity of the solution. Finally, we must be able to infer that, if the agent knows (i) that the test took place, (ii) the observable result of the test, and (iii) how the observable result depends on the unobservable precondition, then he will know the unobservable precondition. Thus we must know enough about knowledge to tell us when an agent's knowing a certain collection of facts implies that he knows other facts as well.

From the preceding discussion, we can conclude that any formalism that enables us to draw inferences about tests at this level of detail must be able to represent the following types of assertions:

(1) After A performs Act, he knows whether Q is true.

(2) After A performs Act, he knows that he has just performed Act.

(3) A knows that Q will be true after he performs *Act* if and only if P is true now.

Moreover, in order to infer what information an agent will gain as a result of performing a test, the formalism must embody, or be able to represent, general principles sufficient to conclude the following:

(4) If (1), (2), and (3) are true, then, after performing *Act*, A will know whether P was true before he performed *Act*.

It is important to emphasize that any work on these problems that is to be of real value must seek to elicit general principles. For instance, it would be possible to represent (1), (2), and (3) in an arbitrary, ad hoc manner and to add an axiom that explicitly states (4), thereby "capturing" the notion of a test. Such an approach, however, would simply restate the superficial observations put forth in this discussion. Our goal in this chapter is to describe a formalism in which specific facts like (4) follow from the most basic principles of reasoning about knowledge and action.

3.2 Formal Theories of Knowledge

A Modal Logic of Knowledge

Since formalisms for reasoning about action have been studied extensively in AI, while formalisms for reasoning about knowledge have not, we will first address the problems of reasoning about knowledge. In Section 3.3 we will see that the formalism that we are led to as a solution to these problems turns out to be well suited to developing an integrated theory of knowledge and action.

The first step in devising a formalism for reasoning about knowledge is to decide what general properties of knowledge we want that formalism to capture. The properties of knowledge in which we will be most interested are those that are relevant to planning and acting. One such property is that anything that is known by someone must be true. If P is false, we would not want to say that anyone knows P. It might be that someone believes P or that someone believes he knows P, but it simply could not be the case that anyone knows P. This is, of course, a major difference between knowledge and belief. If we say that someone believes P, we are not committed to saying that P is either true or false, but if we say that someone knows P, we are committed to the truth of P. The reason that this distinction is important for planning and acting is simply that, for an agent to achieve his goals, the beliefs on which he bases his actions must generally be true. After all, merely believing that performing a certain action will bring about a desired goal is not sufficient for being able to achieve the goal; the action must actually have the intended effect.

Another principle that turns out to be important for planning is that, if someone knows something, he knows that he knows it. This principle is often required for reasoning about plans consisting of several steps. Suppose an agent plans to use Act_1 to achieve his goal, but, in order to perform Act_1 he needs to know whether P is true and whether Q is true. Suppose, further, that he already knows that P is true and that he can find out whether Q is true by performing Act_2. The agent needs to be able to reason that, after performing Act_2, he will know whether P is true and whether Q is true. He knows that he will know whether Q is true because he understands the effects of Act_2, but how does he know that he will know whether P is true? Presumably it works something like this: he knows that P is true, so he knows that he knows that P is true. If he knows how Act_2 affects P, he knows that he will know whether P is true after he performs Act_2. The key step in this argument is an instance of the principle that, if someone knows something, he knows that he know it.

It might seem that we would also want to have the principle that, if someone does not know something, he knows that he does not know it—but this turns out to be false. Suppose that A believes that P, but P is not true. Since P is false, A certainly does not know that P, but it is highly unlikely that he knows that he does not know, since he thinks that P is true.

Probably the most important fact about knowledge that we will want to capture is that agents can reason on the basis of their knowledge. All our examples depend on the assumption that, if an agent trying to solve a problem has all the relevant information, he will apply his knowledge to produce a solution. This creates a difficulty for us, however, since agents (at least human ones) are not, in fact, aware of all the logical consequences of their knowledge. The trouble is that we can never be sure which of the inferences an agent *could* draw, he actually *will*. The principle people normally use in reasoning about what other people know seems to be something like this: if *we* can infer that something is a consequence of what someone knows, then, lacking information to the contrary, we will assume that the other person can draw the same inference.

This suggests the adoption some sort of "default rule" (Reiter 1980) for reasoning about what inferences agents actually draw, but, for the purposes of this study, we will make the simplifying assumption that agents actually do draw all logically valid inferences from their knowledge. We can regard this as the epistemological version of the "frictionless case" in classical physics. For a more general framework in which weaker assumptions about the deductive abilities of agents can be expressed, see the work of Konolige (1985).

Finally, we will need to include the fact that these basic properties of knowledge are themselves *common knowledge*. By this we mean that everyone knows them, and everyone knows that everyone knows them, and everyone knows that everyone knows that everyone knows them, ad infinitum. This type of principle is obviously needed when reasoning about what someone knows about what someone else knows, but it is also important in planning, because an agent must be able to reason about what he will know at various times in the future. In such a case, his "future self" is analogous to another agent.

In his pioneering work on the logic of knowledge and belief, Hintikka (1962) presents a formalism that captures all these properties. We will define a formal logic based on Hintikka's ideas, but modified somewhat to be more compatible with the additional ideas of this chapter. So, what follows is similar to the logic developed by Hintikka in spirit, but not in detail.

The language we will use initially is that of propositional logic, augmented by an operator Know and terms denoting agents. The formula Know(A, P) is interpreted to mean that the agent denoted by the term A knows the proposition expressed by the formula P. So, if John denotes John and Likes(Bill,Mary) means that Bill likes Mary, Know(John,Likes(Bill,Mary)) means that John knows that Bill likes Mary. The axioms of the logic are inductively defined as all instances of the following schemata:

M1. P, such that P is an axiom of ordinary propositional logic
M2. Know(A, P) $\supset P$
M3. Know(A, P) $\supset$ Know(A, Know(A, P))
M4. Know(A, ($P \supset Q$)) $\supset$ (Know(A, P) $\supset$ Know(A, Q))

closed under the principle that

M5. If P is an axiom, then Know(A, P) is an axiom.

The closure of the axioms under the inference rule modus ponens (from ($P \supset Q$) and P, infer Q) defines the theorems of the system. This system is very similar to those studied in modal logic. In fact, if A is held fixed, the resulting system is isomorphic to the modal logic S4 (Hughes and Cresswell 1968). We will refer to this system as the modal logic of knowledge.

These axioms formalize in a straightforward way the principles for reasoning about knowledge that we have discussed. M2 says that anything that is known is true. M3 says that, if someone knows something, he knows that he knows it. M4 says that, if someone knows a formula P and a formula of the form $(P \supset Q)$, then he knows the corresponding formula Q. That is, everyone can (and does) apply modus ponens. M5 guarantees that the axioms are common knowledge. It first applies to M1–M4, which says that everyone knows the basic facts about knowledge; however, since it also applies to its own output, we get axioms stating that everyone knows that everyone knows, etc. Since M5 applies to the axioms of propositional logic (M1), we can infer that everyone knows the facts they represent. Furthermore, because modus ponens is the only inference rule needed in propositional logic, the presence of M4 will enable us to infer that an agent knows any propositional consequence of his knowledge.

A Possible-Worlds Analysis of Knowledge

We could try to use the modal logic of knowledge directly in a computational system for reasoning about knowledge and action, but, as we have argued elsewhere (Moore 1980a), all the obvious ways of doing this encounter difficulties. (Konolige's recent work (1985) suggests some new, more promising possibilities, but some important questions remain to be resolved.) There may well be solutions to these problems, but it turns out that they can be circumvented entirely by changing the language we use to describe what agents know. Instead of talking about the individual propositions that an agent knows, we will talk about what states of affairs are compatible with what he knows. In philosophy, these states of affairs are usually called "possible worlds," so we will adopt that term here as well.

This shift to describing knowledge in terms of possible worlds is based on a rich and elegant formal semantics for systems like our modal logic of knowledge, which was developed by Hintikka (1962, 1971) in his work on knowledge and belief. The advantages of this approach are that it can be formalized within ordinary first-order classical logic in a way that permits the use of standard automatic-deduction techniques in a reasonably efficient manner[2] and that, moreover, it generalizes nicely to an integrated theory for describing the effects of actions on the agent's knowledge.

Possible-world semantics was first developed for the logic of necessity and possibility. From an intuitive standpoint, a possible world may be thought of as a set of circumstances that might have been true

[2]Chapters 6 and 7 of Moore (1980a) present a procedural interpretation of the axioms for knowledge and action given in this chapter that seems to produce reasonably efficient behavior in an automatic deduction system.

in the actual world. Kripke (1963) introduced the idea that a world should be regarded as possible, not absolutely, but only relative to other worlds. That is, the world W_1 might be a possible alternative to W_2, but not to W_3. The relation of one world's being a possible alternative to another is called the *accessibility relation*. Kripke then proved that the differences among some of the most important axiom systems for modal logic corresponded exactly to certain restrictions on the accessibility relation of the possible-world models of those systems. These results are reviewed in Kripke (1971). Concurrently with these developments, Hintikka (1962) published the first of his writings on the logic of knowledge and belief, which included a model theory that resembled Kripke's possible-world semantics. Hintikka's original semantics was done in terms of sets of sentences, which he called *model sets*, rather than possible worlds. Later (Hintikka 1971), however, he recast his semantics using Kripke's concepts, and it is that formulation we will use here.

Kripke's semantics for necessity and possibility can be converted into Hintikka's semantics for knowledge by changing the interpretation of the accessibility relation. To analyze statements of the form $\mathsf{Know}(A, P)$, we will introduce a relation K, such that $K(A, W_1, W_2)$ means that the possible world W_2 is compatible or consistent with what A knows in the possible world W_1. In other words, for all that A knows in W_1, he might just as well be in W_2. It is the set of worlds $\{w_2 \mid K(A, W_1, w_2)\}$ that we will use to characterize what A knows in W_1. We will discuss A's knowledge in W_1 in terms of this set, the set of states of affairs that are consistent with his knowledge in W_1, rather than in terms of the set of propositions he knows. For the present, let us assume that the first argument position of K admits the same set of terms as the first argument position of Know. When we consider quantifiers and equality, we will have to modify this assumption, but it will do for now.

Introducing K is the key move in our analysis of statements about knowledge, so understanding what K means is particularly important. To illustrate, suppose that in the actual world—call it W_0—A knows that P, but does not know whether Q. If W_1 is a world where P is false, then W_1 is not compatible with what A knows in W_0; hence we would have $\neg K(A, W_0, W_1)$. Suppose that W_2 and W_3 are compatible with everything A knows, but that Q is true in W_2 and false in W_3. Since A does not know whether Q is true, for all he knows, he might be in either W_2 or W_3 instead of W_0. Hence, we would have both $K(A, W_0, W_2)$ an $K(A, W_0, W_3)$. This is depicted graphically in Figure 1.

Some of the properties of knowledge can be captured by putting

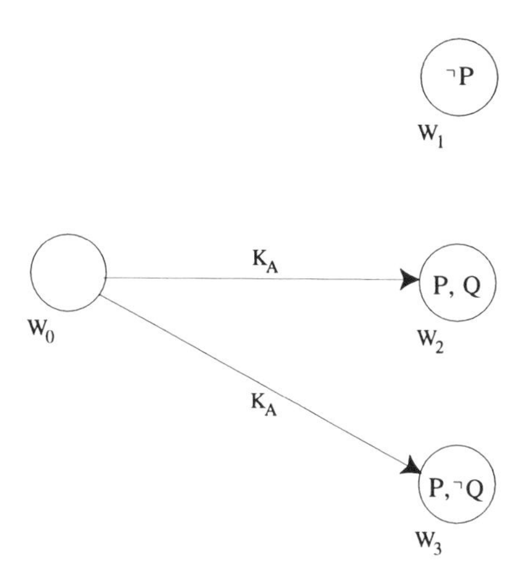

Figure 1. "A knows that P"
"A doesn't know whether Q"

constraints on the accessibility relation K. For instance, requiring that the actual world W_0 be compatible with what each knower knows in W_0, i.e., $\forall a_1(K(a_1, W_0, W_0))$, is equivalent to saying that anything that is known is true. That is, if the actual world is compatible with what everyone [actually] knows, then no one has any false knowledge. This corresponds to the modal axiom M2.

The definition of K implies that, if A knows that P in W_0, then P must be true in every world W_1 such that $K(A, W_0, W_1)$. To capture the fact that agents can reason with their knowledge, we will assume the converse is also true. That is, we assume that, if P is true in every world W_1 such that $K(A, W_0, W_1)$, then A knows that P in W_0. (See Figure 2.) This principle is the model-theoretic analogue of axiom M4 in the modal logic of knowledge. To see that this is so, suppose that A knows that P and that $(P \supset Q)$. Therefore, P and $(P \supset Q)$ are both true in every world that is compatible with what A knows. If this is the case, though, then Q must be true in every world that is compatible with what A knows. By our assumption, therefore, we conclude that A knows that Q.

Since this assumption, like M4, is equivalent to saying that an agent knows all the logical consequences of his knowledge, it should be interpreted only as a default rule. In a particular instance, the fact that P follows from A's knowledge would be a justification for concluding that

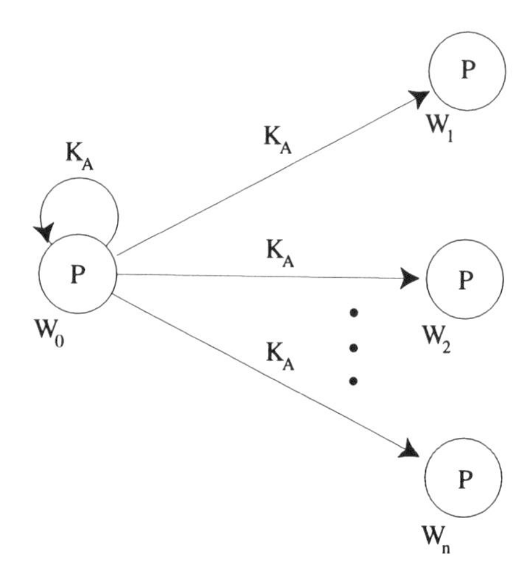

Figure 2. "P is true in every world that is compatible with what A knows"

A knows P. However, we should be prepared to retract the conclusion that A knows P in the face of stronger evidence to the contrary.

With this assumption, we can get the effect of M3—the axiom stating that, if someone knows something, he knows that he knows it—by requiring that, for any W_1 and W_2, if W_1 is compatible with what A knows in W_0 and W_2 is compatible with what A knows in W_1, then W_2 is compatible with what A knows in W_0. Formally expressed, this is

$$\forall a_1, w_1, w_2(K(a_1, W_0, w_1) \supset (K(a_1, w_1, w_2) \supset K(a_1, W_0, w_2)))$$

By our previous assumption, the facts that A knows are those that are true in every world that is compatible with what A knows in the actual world. Furthermore, the facts that A knows that he knows are those that are true in every world that is compatible with what he knows in every world that is compatible with what he knows in the actual world. By the constraint we have just proposed, however, all these worlds must also be compatible with what A knows in the actual world (see Figure 3), so, if A knows that P, he knows that he knows that P.

Finally, we can get the effect of M5, the principle that the basic facts about knowledge are themselves common knowledge, by generalizing these constraints so that they hold not only for the actual world, but for all possible worlds. This follows from the fact that, if these constraints hold for all worlds, they hold for all worlds that are com-

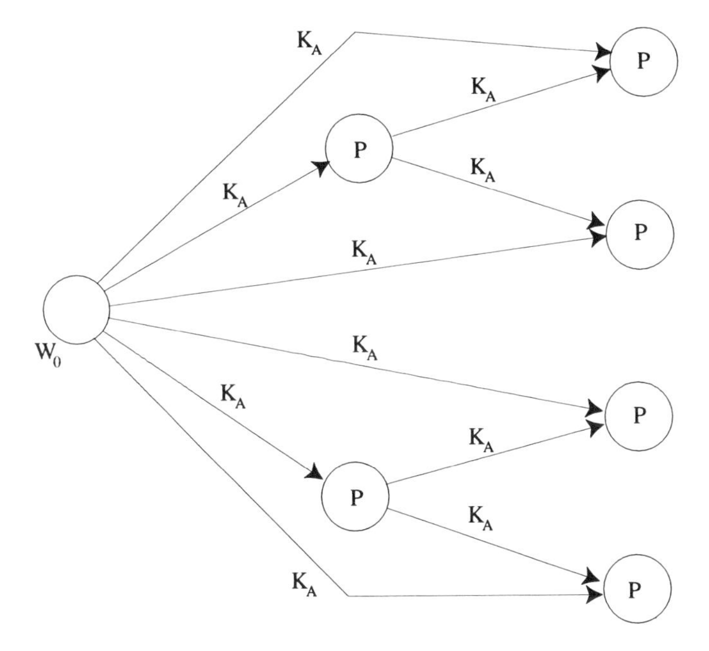

Figure 3. "If A knows that P, then he knows that he knows that P"

patible with what anyone knows in the actual world; they also hold for all worlds that are compatible with what anyone knows in all worlds that are compatible with what anyone knows in the actual world, etc. Therefore, everyone knows the facts about knowledge that are represented by the constraints, and everyone knows that everyone knows, etc. Note that this generalization has the effect that the constraint corresponding to M2 becomes the requirement that, for a given knower, K is reflexive, while the constraint corresponding to M3 becomes the requirement that, for a given knower, K is transitive.

Analyzing knowledge in terms of possible worlds gives us a very nice treatment of knowledge about knowledge. Suppose A knows that B knows that P. Then, if the actual world is W_0, in any world W_1 such that $K(A, W_0, W_1)$, B knows that P. We now continue the analysis relative to W_1, so that, in any world W_2 such that $K(B, W_1, W_2)$, P is true. Putting both stages together, we obtain the analysis that, for any worlds W_1 and W_2 such that $K(A, W_0, W_1)$ and $K(B, W_1, W_2)$, P is true in W_2 (See Figure 4.)

Given these constraints and assumptions, whenever we want to assert or deduce something that would be expressed in the modal logic

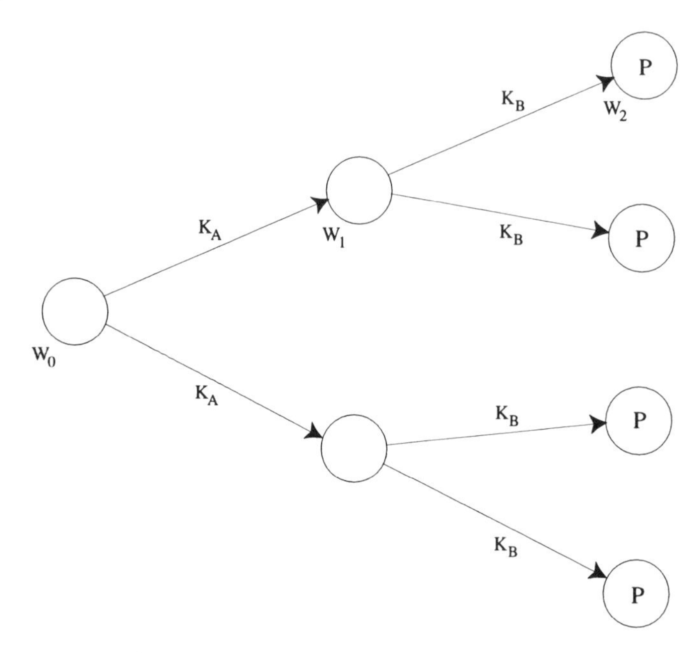

Figure 4. "A knows that B knows that P"

of knowledge by Know(A, P), we can instead assert or deduce that P is true in every world that is compatible with what A knows. We can express this in ordinary first-order logic, by treating possible worlds as individuals (in the logical sense), so that K is just an ordinary relation. We will therefore introduce an operator T such that $T(W, P)$ means that the formula P is true in the possible world W. If we let W_0 denote the actual world, we can convert the assertion Know(A, P) into

$$\forall w_1(K(A, W_0, w_1) \supset T(w_1, P))$$

It may seem that we have not made any real progress, since, although we have gotten rid of one nonstandard operator, Know, we have introduced another one, T. However, T has an important property that Know does not. Namely, T "distributes" over ordinary logical operators. In other words, $\neg P$ is true in W just in case P is not true in W, $(P \vee Q)$ is true in W just in case P is true in W or Q is true in W, and so on. We might say that T is extensional, relative to a possible world. This means that we can transform any formula so that T is applied only to atomic formulas. We can then turn T into an ordinary first-order relation by treating all the atomic formulas as *names* of atomic propositions, or we can get rid of T by replacing the atomic formulas with predicates on possible worlds. This is no loss to the expressive power of the language, since, where we would have previously asserted P, we now simply assert $T(W_0, P)$ or $P(W_0)$ instead.

Knowledge, Equality, and Quantification

The formalization of knowledge presented so far is purely propositional; a number of additional problems arise when we attempt to extend the theory to handle equality and quantification. For instance, as Frege (1949) pointed out, attibutions of knowledge and belief lead to violations of the principle of equality substitution. We are not entitled to infer $\mathsf{Know}(A, P(C))$ from $B = C$ and $\mathsf{Know}(A, P(B))$ because A might not know that the identity holds.

The possible-world analysis of knowledge provides a very neat solution to this problem, once we realize that a term can denote different objects in different possible worlds. For instance, if B is the expression "the number of planets" and C is "nine," then, although $B = C$ is true in the actual world, it would be false in a world in which there was a tenth planet. Thus, we will say that an equality statement such as $B = C$ is true in a possible world W just in case the denotation of the term B in W is the same as the denotation of the term C in W. This is a special case of the more general rule that a formula of the form $P(A_1, \ldots, A_n)$ is true in W just in case the tuple consisting of the denotations in W of the terms $A_1, \ldots, A_n$ is in the extension in W of the relation expressed by P, provided that we fix the interpretation of $=$ in all possible worlds to be the identity relation.

Given this interpretation, the inference of $\mathsf{Know}(A, P(C))$ from $B = C$ and $\mathsf{Know}(A, P(B))$ will be blocked (as it should be). To infer $\mathsf{Know}(A, P(C))$ from $\mathsf{Know}(A, P(B))$ by identity substitution, we would have to know that B and C denote the same object in *every* world compatible with what A knows, but the truth of $B = C$ guarantees only that they denote the same object in the actual world. On the other hand, if $\mathsf{Know}(A, P(B))$ and $\mathsf{Know}(A, (B = C))$ are both true, then in all worlds that are compatible with what A knows, the denotation of B is in the extension of P and is the same as the denotation of C; hence, the denotation of C is in the extension of P. From this we can infer that $\mathsf{Know}(A, P(C))$ is true.

The introduction of quantifiers also causes problems. To modify a famous example from Quine (1971b), consider the sentence "Ralph knows that someone is a spy." This sentence has at least two interpretations. One is that Ralph knows that there is at least one person who is a spy, although he may have no idea who that person is. The other interpretation is that there is a particular person whom Ralph knows to be a spy. As Quine says (1971b, p. 102), "The difference is vast; indeed, if Ralph is like most of us, [the first] is true and [the second] is false." This ambiguity was explained by Russell (1949) as a difference of *scope*. The idea is that indefinite noun phrases such as "someone" can be analyzed in context by paraphrasing sentences of the form $P(\text{"someone"})$ as "There exists a person x such that $P(x)$," or, more formally, $\exists x(\mathsf{Person}(x) \wedge P(x))$. Russell goes on to point out that, in sentences of the form "A knows that someone is a P," the rule for eliminating "someone" can be applied to either the whole sentence or only the subordinate clause, "someone is a P." Applying this observation to "Ralph knows that someone is a spy," gives us the following two formal representations:

(1) $\mathsf{Know}(\mathsf{Ralph}, \exists x(\mathsf{Person}(x) \wedge \mathsf{Spy}(x)))$

(2) $\exists x(\mathsf{Person}(x) \wedge \mathsf{Know}(\mathsf{Ralph}, \mathsf{Spy}(x)))$

The most natural English paraphrases of these formulas are "Ralph knows that there is a person who is a spy," and "There is a person who Ralph knows is a spy." These seem to correspond pretty well to the two interpretations of the original sentence. So, the ambiguity in the original sentence is mapped into an uncertainty as to the scope of the operator Know relative to the existential quantifier introduced by the indefinite description "someone."

Following a suggestion of Hintikka (1962), we can use a formula similar to (2) to express the fact that someone knows who or what something is. He points out that a sentence of the form "A knows who (or what) B is" intuitively seems to be equivalent to "there is someone (or something) that A knows to be B." But this can be represented formally as $\exists x(\mathsf{Know}(A, (x = B)))$. To take a specific example, "John knows who the President is" can be paraphrased as "There is someone whom John knows to be the President," which can be represented by

(3) $\exists x(\mathsf{Know}(\mathsf{John}, (x = \mathsf{President})))$

In (1), Know may still be regarded as a purely propositional operator, although the proposition to which it is applied now has a quantifier in it. Put another way, Know still is used simply to express a relation between a knower and the proposition he knows. But (2) and (3) are not so simple. In these formulas there is a quantified variable that, although bound outside the scope of the operator Know, has an occurrence inside; this is sometimes called "quantifying in." Quantifying into knowledge and belief contexts is frequently held to pose serious problems of interpretation. Quine (1971b), for instance, holds that it is unintelligible, because we have not specified what proposition is known unless we say what description is used to fix the value of the quantified variable.

The possible-world analysis, however, provides us with a very natural interpretation of quantifying in. We keep the standard interpretation that $\exists x(P(x))$ is true just in case there is some value for x that satisfies P. If P is $\mathsf{Know}(A, Q(x))$, then a value for x satisfies $P(x)$ just in case that value satisfies $Q(x)$ in every world that is compatible with what A knows. So (2) is satisfied if there is a particular person who is a spy in every world that is compatible with what A knows. That is, in every such world the same person is a spy. On the other hand, (1) is satisfied if, in every world compatible with what A knows, there is some person who is a spy, but it does not have to be the same one in each case.

Note that the difference between (1) and (2) has been transformed from a difference in the relative scopes of an existential quantifier and the operator Know to a difference in the relative scopes of an existential and a universal quantifier (the "every" in "every possible world compatible with...") Recall from ordinary first-order logic that $\exists x(\forall y(P(x, y)))$ entails $\forall y(\exists x(P(x, y)))$, but not vice versa. The possible-world analysis, then, implies that we should be able to infer "Ralph knows that there is a spy," from "There is someone Ralph knows to be a spy," as indeed we can.

When we look at how this analysis applies to our representation for "knowing who," we get a particularly satisfying picture. We said that A knows who B is means that there is someone whom A knows to be B. If we analyze this, we conclude that there is a particular individual who is B in every world that is compatible with what A knows. Suppose this were not the case, and that, in some of the worlds compatible with what A knows, one person is B, whereas in the other worlds, some other person is B. In other words, for all that A knows, either of these two people might be B. But this is exactly what we mean when we say that A does not know who B is! Basically, the possible-world view gives us the very natural picture that A knows who B is if A has narrowed the possibilities for B down to a single individual.

Another consequence of this analysis worth noting is that, if A knows who B is and A knows who C is, we can conclude that A knows whether $B = C$. If A knows who B is and who C is, then B has the the same denotation in all the worlds that are compatible with what A knows, and this is also true for C. Since, in all these worlds, B and C each have only one denotation, they either denote the same thing everywhere or denote different things everywhere. Thus, either $B = C$

is true in every world compatible with what A knows or $B \neq C$ is. From this we can infer that either A knows that B and C are the same individual or that they are not.

We now have a coherent account of quantifying in that is not framed in terms of knowing particular propositions. Still, in some cases knowing a certain proposition counts as knowing something that would be expressed by quantifying in. For instance, the proposition that John knows that 321-1234 is Bill's telephone number might be represented as

(4) Know(John,(321-1234 = Phone-Num(Bill))),

which does not involve quantifying in. We would want to be able to infer from this, however, that John knows what Bill's telephone number is, which would be represented as

(5) $\exists x$(Know(John, (x = Phone-Num(Bill)))).

It might seem that (5) can be derived from (4) simply by the logical principle of existential generalization, but that principle is not always valid in knowledge contexts. Suppose that (4) were not true, but that instead John simply knew that Mary and Bill had the same telephone number. We could represent this as

(6) Know(John, (Phone-Num(Mary) = Phone-Num(Bill))).

It is clear that we would not want to infer from (6) that John knows what Bill's telephone number is—yet, if existential generalization were universally valid in knowledge contexts, this inference would be valid.

It therefore seems that, in knowledge contexts, existential generalization can be applied to some referring expressions ("321-1234"), but not to others ("Mary's telephone number"). We will call the expressions to which existential generalization can be applied *standard identifiers*, since they seem to be the ones an agent would use to identify an object for another agent. That is, "321-1234" is the kind of answer that would *always* be appropriate for telling someone what John's telephone number is, whereas "Mary's telephone number," as a general rule, would not.[3]

In terms of possible worlds, standard identifiers have a very straightforward interpretation. Standard identifiers are simply terms that have

[3] "Mary's telephone number" *would* be an appropriate way of telling someone what John's telephone number was if he already knew Mary's telephone number, but this knowledge would consist in knowing what expression of the type "321-1234" denoted Mary's telephone number. Therefore, even in this case, using "Mary's telephone number" to identify John's telephone number would just be an indirect way of getting to the standard indentifier.

the same denotation in every possible world. Following Kripke (1972), we will call terms that have the same denotation in every possible world *rigid designators*. The conclusion that standard identifiers are rigid designators seems inescapable. If a particular expression can always be used by an agent to identify its referent for any other agent, then there must not be any possible circumstances under which it could refer to something else. Otherwise, the first agent could not be sure that the second was in a position to rule out those other possibilities.

The validity of existential generalization for standard identifiers follows immediately from their identification with rigid designators. The possible-world analysis of $\mathsf{Know}(A, P(B))$ is that, in every world compatible with what A knows, the denotation of B in that world is in the extension of P in that world. Existential generalization fails in general because we are unable to conclude that there is any particular individual that is in the extension of P in all the relevant worlds. If B is a rigid designator, however, the denotation of B is the same in every world. Consequently, it is the same in every world compatible with what A knows, and that denotation is an individual that is in the extension of P in all those worlds.

There are a few more observations to be made about standard identifiers and rigid designators. First, in describing standard identifiers we assumed that everyone knew what they referred to. Identifying them with rigid designators makes the stronger claim that what they refer to is common knowledge. That is, not only does everyone know what a particular standard identifier denotes, but everyone knows that everyone knows, etc. Second, although it is natural to think of any individual having a unique standard identifier, this is not required by our theory. What the theory does require is that, if there are two standard identifiers for the same individual, it should be common knowledge that they denote the same individual.

3.3 Formalizing the Possible-World Analysis of Knowledge

Object Language and Metalanguage

As we indicated above, the analysis of knowledge in terms of possible worlds can be formalized completely within first-order logic by admitting possible worlds into the domain of quantification and making the extension of every expression depend on the possible world in which it is evaluated. For example, the possible-world analysis of "A knows who B is" would be as follows: There is some individual x such that, in every world w_1 that is compatible with what the agent who is A in the actual world knows in the actual world, x is B in w_1. This means that in our formal theory we translate the formula of the modal logic of knowledge,

$$\exists x(\mathsf{Know}(A,(x = B))),$$

into the first-order formula,

$$\exists x(\forall w_1(K(A(W_0), W_0, w_1) \supset (x = B(w_1)))).$$

One convenient way of stating the translation rules precisely is to axiomatize them in our first-order theory of knowledge. This can be done by introducing terms to denote formulas of the modal logic of knowledge (which we will henceforth call the *object language*) and axiomatizing a truth definition for those formulas in a first-order language that talks about possible worlds (the *metalanguage*). This has the advantage of letting us use either the modal language or the possible-world language—whichever is more convenient for a particular purpose—while rigorously defining the connection between the two.

The typical method of representing expressions of one formal language in another is to use string operations like concatenation or list operations like Cons in LISP, so that the conjunction of P and Q might be represented by something like $\mathsf{Cons}(P, \mathsf{Cons}('\wedge, \mathsf{Cons}(Q, \mathsf{Nil})))$, which could be abbreviated $\mathsf{List}(P,'\wedge, Q)$. This would be interpreted as a list whose elements are P followed by the conjunction symbol followed by Q. Thus, the metalanguage expression $\mathsf{Cons}(P, \mathsf{Cons}('\wedge, \mathsf{Cons}(Q, \mathsf{Nil})))$ would denote the object language expression $(P \wedge Q)$. McCarthy (1962) has devised a much more elegant way to do the encoding, however. For purposes of semantic interpretation of the object language, which is what we want to do, the details of the syntax of that language are largely irrelevant. In particular, the only thing we need to know about the syntax of conjunctions is that there is *some* way of taking P and Q and producing the conjunction of P and Q. We can represent this by having a function And such that $\mathsf{And}(P, Q)$ denotes the conjunction of P and Q. To use McCarthy's term, $\mathsf{And}(P, Q)$ is an *abstract syntax* for representing the conjunction of P and Q.

We will represent object language variables and constants by metalanguage constants; we will use metalanguage functions in an abstract syntax to represent object language predicates, functions, and sentence operators. For example, we will represent the object language formula $\mathsf{Know}(\mathsf{John}, \exists x(P(x)))$ by the metalanguage term Know (John,Exist(X,P(X))), where John and X are metalanguage constants, and Know, Exist, and P are metalanguage functions.

Since Know(John,Exist(X,P(X))) is a term, if we want to say that the object language formula it denotes is true, we have to do so explicitly by means of a metalanguage predicate True:

True(Know(John,Exist(X,P(X)))).

In the possible-world analysis of statements about knowledge, however, an object language formula is not absolutely true, but only relative to a possible world. Hence, True expresses not absolute truth, but truth in the actual world, which we will denote by W_0. Thus, our first axiom is

L1. $\forall p_1(\mathsf{True}(p_1) \equiv T(W_0, p_1))$,

where $T(W, P)$ means that formula P is true in world W. To simplify the axioms, we will let the metalanguage be a many-sorted logic, with different sorts assigned to differents sets of variables. For instance, the variables w_1, w_2,... will range over possible worlds; x_1, x_2,... will range over individuals in the domain of the object language; and a_1, a_2,... will range over agents. Because we are axiomatizing the object language itself, we will need several sorts for different types of object language expressions. The variables p_1, p_2,... will range over object language formulas, and t_1,t_2,... will range over object language terms.

The recursive definition of T for the propositional part of the object language is as follows:

L2. $\forall w_1, p_1, p_2(T(w_1, \mathsf{And}(p_1, p_2)) \equiv (T(w_1, p_1) \wedge T(w_1, p_2)))$
L3. $\forall w_1, p_1, p_2(T(w_1, \mathsf{Or}(p_1, p_2)) \equiv (T(w_1, p_1) \vee T(w_1, p_2)))$
L4. $\forall w_1, p_1, p_2(T(w_1, \mathsf{Imp}(p_1, p_2)) \equiv (T(w_1, p_1) \supset T(w_1, p_2)))$
L5. $\forall w_1, p_1, p_2(T(w_1, \mathsf{Iff}(p_1, p_2)) \equiv (T(w_1, p_1) \equiv T(w_1, p_2)))$
L6. $\forall w_1, p_1(T(w_1, \mathsf{Not}(p_1)) \equiv \neg T(w_1, p_1))$

Axioms L1–L6 merely translate the logical connectives from the object language to the metalanguage, using an ordinary Tarskian truth definition. For instance, according to L2, And(P, Q) is true in a world if and only if P and Q are both true in the world. The other axioms state that all the truth-functional connectives are "transparent" to T in exactly the same way.

To represent quantified object language formulas in the metalanguage, we will introduce additional functions into the abstract syntax: Exist and All. These functions will take two arguments—a term denoting an object language variable and a term denoting an object language formula. Axiomatizing the interpretation of quantified object language formulas presents some minor technical problems, however. We would like to say something like this: Exist(X, P) is true in W if and only if there is some individual such that the open formula P is true of that individual in W. We do not have any way of saying that an open formula is true of an individual in a world, however; we just have the predicate T, which simply says that a formula is true in a world. One way of solving the problem would be to introduce a new predicate, or perhaps redefine T, to express the Tarskian notion of satisfaction rather than truth. This approach is semantically clean but syntactically clumsy, so we will instead follow the advice of Scott (1970, p. 151) and define the truth of a quantified statement in terms of substituting into the body of that statement a rigid designator for the value of the quantified variable.

In order to formalize this substitutional approach to the interpretation of object language quantification, we need a rigid designator in the object language for every individual. Since our representation of the object language is in the form of an abstract syntax, we can simply stipulate that there is a function @ that maps any individual in the object language's domain of discourse into an object language rigid designator of that individual. The definition of T for quantified statements is then given by the following axiom schemata:

L7. $\forall w_1(T(w_1, \mathsf{Exist}(X, P)) \equiv \exists x_1(T(w_1, P[@(x_1)/X])))$

L8. $\forall w_1(T(w_1, \mathsf{All}(X, P)) \equiv \forall x_1(T(w_1, P[@(x_1)/X])))$

In these schemata, P may be any object language formula, X may be any object language variable, and the notation $P[@(x_1)/X]$ designates the expression that results from substituting $@(x_1)$ for every free occurrence of X in P.

L7 says that an existentially quantified formula is true in a world W if and only if, for *some* individual, the result of substituting a rigid designator of that individual for the bound variable in the body of the formula is true in W. L8 says that a universally quantified formula is true in W if and only if, for *every* individual, the result of substituting a rigid designator of that individual for the bound variable in the body of the formula is true in W.

Except for the knowledge operator itself, the only part of the truth definition of the object language that remains to be given is the definition of T for atomic formulas. We remarked previously that a formula of the form $P(A_1, \ldots, A_n)$ is true in a world W just in case the tuple consisting of the denotations in W of the terms $A_1, \ldots, A_n$ is in the extension in W of the relation P. To axiomatize this principle, we need two additions to the metalanguage. First, we need a function D that maps a possible world and an object language term into the denotation of that term in that world. Second, for each n-place object language predicate P, we need a corresponding n+1-place metalanguage predicate (which, by convention, we will write $:P$) that takes as its arguments the possible world in which the object language formula is to be evaluated and the denotations in that world of the arguments of the object language predicate. The interpretation of an object language atomic formula is then given by the axiom schema

L9. $\forall w_1, t_1, \ldots, t_n$
$$(T(w_1, P(t_1, \ldots, t_n)) \equiv :P(w_{,1}, D(w_1, t_1), \ldots, D(w_1, t_n)))$$

To eliminate the function D, we need to introduce a metalanguage expression corresponding to each object language constant or function. In the general case, the new expression will be a function with an extra argument position for the possible world of evaluation. The axiom schemata for D are then

L10. $\forall w_1, x_1(D(w_1, @(x_1)) = x_1)$
L11. $\forall w_1(D(w_1, C) = :C(w_1))$
L12. $\forall w_1, t_1, \ldots, t_n$
$$(D(w_1, F(t_1, \ldots, t_n)) = :F(w_1, D(w_1, t_1), \ldots, D(w_1, t_n))),$$

where C is an object language constant and F is an object language function, and we use the ":" convention already introduced for their metalanguage counterparts.

Since $@(x_1)$ is a rigid designator of x_1, its value is x_1 in every possible world. In the general case, an object language constant will have a corresponding metalanguage function that picks out the denotation of the constant in a particular world. Similarly, an object language function will have a corresponding metalanguage function that maps a possible world and the denotations of the arguments of the object language function into the value of the object language function applied to those arguments in that world.

It will be convenient to treat specially those object language constants and functions that are (or can be used to construct) rigid designators. We could introduce additional axioms asserting that such expressions have the same value in every possible world, but we can accomplish the same end simply by making the corresponding metalanguage expressions independent of the possible world of evaluation. So, for object language constants that are rigid designators, we will have a variant of axiom L11:

L11a. $\forall w_1(D(w_1, C) = :C)$ if C is a rigid designator.

We will similarly treat *rigid functions*—those that always map a particular tuple of arguments into the same value in all possible worlds:

L12a. $\forall w_1, t_1, \ldots, t_n$
$$(D(w_1, F(t_1, \ldots, t_n)) = {:}F(D(w_1, t_1), \ldots, D(w_1, t_n)))$$
if F is a rigid function.

Finally, we introduce a special axiom for the equality predicate of the object language, fixing its interpretation in all possible worlds to be the identity relation:

L13. $\forall w_1, t_1, t_2(T(w_1, \mathsf{Eq}(t_1, t_2)) \equiv (D(w_1, t_1) = D(w_1, t_2)))$

A First-Order Theory of Knowledge

The axioms given in the preceding section allow us to talk about a formula of first-order logic being true relative to a possible world rather than absolutely. This generalization would be pointless, however, if we never had occasion to mention any possible worlds other than the actual one. References to other possible worlds are introduced by our axioms for knowledge:

K1. $\forall w_1, t_1, p_1$
$$(T(w_1, \mathsf{Know}(t_1, p_1)) \equiv \forall w_2(K(D(w_1, t_1), w_1, w_2) \supset T(w_2, p_1)))$$

K2. $\forall a_1, w_1(K(a_1, w_1, w_1))$

K3. $\forall a_1, w_1, w_2$
$$(K(a_1, w_1, w_2) \supset \forall w_3(K(a_1, w_2, w_3) \supset K(a_1, w_1, w_3)))$$

K1 gives the possible-world analysis for object language formulas of the form $\mathsf{Know}(A, P)$. The interpretation is that $\mathsf{Know}(A, P)$ is true in world W_1 just in case P is true in every world that is compatible with what the agent denoted by A in W_1 knows in W_1. Since an object language term may denote different individuals in different possible worlds, we use $D(W_1, A)$ to identify the denotation of A in W_1. K represents the accessibility relation associated with Know, so $K(D(W_1, A), W_1, W_2)$ is how we represent the fact W_2 is compatible with what the agent denoted by A in W_1 knows in W_1.

As we pointed out before, the principle embodied in K1 is that an agent knows everything entailed by his knowledge. Since this is too strong a generalization, in a more thorough analysis we would regard the inference from the right side of K1 to the left side as being a default inference. K2 and K3 state constraints on the accessibility relation K that we use to capture other properties of knowledge. They require that, for a fixed agent $:A$, $K(:A, w_1, w_2)$ be reflexive and transitive. We have already shown this entails that anything that anyone knows must be true, and that if someone knows something he knows that he knows it. Finally, the fact that K1–K3 are asserted to hold for all possible worlds implies that everyone knows the principles they embody, and everyone knows that everyone knows, etc. In other words, these principles are common knowledge.

To illustrate how our theory operates, we will show how to derive a simple result in the logic of knowledge, that from the premises that A knows that $P(B)$ and A knows that $B = C$, we can conclude that A knows that $P(C)$. Our proofs will be in natural-deduction form. The axioms and preceding lines that justify each step will be given to the right of the step. Subordinate proofs will be indicated by indented sections, and ASS will mark the assumptions on which these subordinate proofs are based. DIS(N, M) will indicate the discharge of the assumption on line N with respect to the conclusion on line M. The general pattern of proofs in this system will be to assert the object language premises of the problem, transform them into their metalanguage equivalents, using axioms L1–L13 and K1, then derive the metalanguage version of the conclusion using first-order logic and axioms such as K2 and K3, and finally transform the conclusion back into the object language, again using L1–L13 and K1.

Given: True(Know(A,P(B)))
       True(Know(A,Eq(B,C)))

Prove: True(Know(A,P(C)))

1.	$\mathsf{True(Know(A, P(B)))}$	Given
2.	$T(W_0, \mathsf{Know(A, P(B))})$	L1,1
3.	$K(D(W_0, A), W_0, w_1) \supset T(w_1, \mathsf{P(B)})$	K1,2
4.	$K(:A(W_0), W_0, w_1) \supset T(w_1, \mathsf{P(B)})$	L11,3
5.	$\mathsf{True(Know(A, Eq(B, C)))}$	Given
6.	$T(W_0, \mathsf{Know(A, Eq(B, C))})$	L1,5
7.	$K(D(W_0, \mathsf{A}), W_0, w_1) \supset T(w_1, \mathsf{Eq(B, C)})$	K1,6
8.	$K(:A(W_0), W_0, w_1) \supset T(w_1, \mathsf{Eq(B, C)})$	L11,7
9.	$\quad K(:A(W_0), W_0, w_1)$	ASS
10.	$\quad T(w_1, \mathsf{P(B)})$	4,9
11.	$\quad :P(w_1, \mathsf{D}(w_1, \mathsf{B}))$	L9,10
12.	$\quad :P(w_1, :B(w_1))$	L11,11
13.	$\quad T(w_1, \mathsf{Eq(B, C)})$	8,9
14.	$\quad D(w_1, B) = D(w_1, \mathsf{C})$	L13,13
15.	$\quad :B(w_1) = :C(w_1)$	L11,14
16.	$\quad :P(w_1, :C(w_1))$	12,15
17.	$\quad :P(w_1, D(w_1, \mathsf{C}))$	L11,16
18.	$\quad T(w_1, \mathsf{P(C)})$	L9,17
19.	$K(:A(W_0), W_0, w_1) \supset T(w_1, \mathsf{P(C)})$	DIS(9,18)
20.	$K(D(W_0, A), W_0, w_1) \supset T(w_1, \mathsf{P(C)})$	L11,19

21. $T(W_0, \mathsf{Know}(\mathsf{A}, \mathsf{P}(\mathsf{C})))$ K1,20
22. True(Know(A, P(C))) L1,21

A knows that $P(B)$ (Line 1), so $P(B)$ is true in every world compatible with what A knows (Line 4). Similarly, since A knows that $B = C$ (Line 5), $B = C$ is true in every world compatible with what A knows (Line 8). Let w_1 be one of these worlds (Line 9). $P(B)$ and $B = C$ must be true in w_1 (Lines 12 and 15), hence $P(C)$ must be true in w_1 (Line 16). Therefore, $P(C)$ is true in every world compatible with what A knows (Line 19), so A knows that $P(C)$ (Line 22). If True(Eq(B,C)) had been given instead of True(Know(A,Eq(B,C))), we would have had $B = C$ true in W_0 instead of w_1. In that case, the substitution of C for B in $P(B)$ (Line 16) would not have been valid, and we could not have concluded that A knows that $P(C)$. This proof seems long because we have made each routine step a separate line. This is worth doing once to illustrate all the formal details, but in subsequent examples we will combine some of the routine steps to shorten the derivation.

3.4 A Possible-Worlds Analysis of Action

In the preceding sections, we have presented a framework for describing what someone knows in terms of possible worlds. To characterize the relation of knowledge to action, we need a theory of action in these same terms. Fortunately, the standard way of looking at actions in AI gives us just that sort of theory. Most AI programs that reason about actions are based on a view of the world as a set of possible states of affairs, with each action determining a binary relation between states of affairs—one being the outcome of performing the action in the other. We can integrate our analysis of knowledge with this view of action by identifying the possible worlds used to describe knowledge with the possible states of affairs used to describe actions.

The identification of a possible world, as used in the analysis of knowledge, with the state of affairs at a particular time does not require any changes in the formalization already presented, but it does require a reinterpretation of what the axioms mean. If the variables w_1, w_2,...are reinterpreted as ranging over states of affairs, then "A knows that P" will be analyzed roughly as "P is true in every state of affairs that is compatible with what A knows in the actual state of affairs." It might seem that taking possible worlds to be states of affairs, and therefore not extended in time, might make it difficult to talk about what someone knows regarding the past or future. That is not the case, however. Knowledge about the past and future can be handled by modal tense operators, with corresponding accessibility relations between possible states-of-affairs/worlds. We could have a tense operator Future such that Future(P) means that P will be true at some time to come. If we let F be an accessibility relation such that $F(W_1, W_2)$ means that the state-of-affairs/world W_2 lies in the future of the state-of-affairs/world W_1, then we can define Future(P) to be true in W_1 just in case there is some W_2 such that $F(W_1, W_2)$ holds and P is true in W_2.

This much is standard tense logic (e.g., Rescher and Urquhart 1971). The interesting point is that statements about someone's knowledge of the future work out correctly, even though such knowledge is analyzed in terms of alternatives to a state of affairs, rather than alternatives to a possible world containing an entire course of events. The proposition that John knows that P will be true is represented simply by Know(John, Future(P)). The analysis of this is that Future(P) is true in every state of affairs that is compatible with what John knows, from which it follows that, for each state of affairs that is compatible with what John knows, P is true in some future alternative to that state of affairs. An important point to note here is that two states of affairs can be "internally" similar (that is, they coincide in the truth-value assigned to any nonmodal statement), yet be distinct because they differ in the accessibility relations they bear to other possible states of affairs. Thus, although we treat a possible world as a state of affairs rather than a course of events, it is a state of affairs in the particular course of events defined by its relationships to other states of affairs.

For planning and reasoning about future actions, instead of a tense operator like Future, which simply asserts what will be true, we need an operator that describes what would be true if a certain event occurred. Our approach will be to recast McCarthy's situation calculus (McCarthy 1968, McCarthy and Hayes 1969) so that it meshes with our possible-world characterization of knowledge. The situation calculus is a first-order language in which predicates that can vary in truth-value over time are given an extra argument to indicate what situations (i.e., states of affairs) they hold in, with a function Result that maps an agent, an action, and a situation into the situation that results from the agent's performance of the action in the first situation. Statements about the effects of actions are then expressed by formulas like P(Result(A,Act,S)), which means that P is true in the situation that results from A's performing Act in situation S.

To integrate these ideas into our logic of knowledge, we will reconstruct the situation calculus as a modal logic. In parallel to the operator Know for talking about knowledge, we introduce an object language operator Res for talking about the results of events. Situations will not be referred to explicitly in the object language, but they will reappear in the possible-world semantics for Res in the metalanguage. Res will be a two-place operator whose first argument is a term denoting an event, and whose second argument is a formula. $\mathsf{Res}(E, P)$ will mean that it is possible for the event denoted by E to occur and that, if it did, the formula P would then be true. The possible-world semantics for Res will be specified in terms of an accessiblity relation R, parallel to K, such that $R(:E, W_1, W_2)$ means that W_2 is the situation/world that would result from the event $:E$ happening in W_1.

We assume that, if it is impossible for $:E$ to happen in W_1 (i.e., if the prerequisites of $:E$ are not satisfied), then there is no W_2 such that $R(:E, W_1, W_2)$ holds. Otherwise we assume that there is exactly one W_2 such that $(:E, W_1, W_2)$ holds[4]

R1. $\forall w_1, w_2, w_3, e_1((R(e_1, w_1, w_2) \wedge R(e_1, w_1, w_3)) \supset (w_2 = w_3))$

(Variables e_1, e_2,...range over events.) Given these assumptions, $\mathsf{Res}(E, P)$ will be true in a situation/world W_1 just in case there is some W_2 that is the situation/world that results from the event described by E happening in W_1, and in which P is true:

R2. $\forall w_1, t_1, p_1$
$$(T(w_1, \mathsf{Res}(t_1, p_1)) \equiv \exists w_2(R(D(w_1, t_1), w_1, w_2) \wedge T(w_2, p_1)))$$

The type of event we will normally be concerned with is the performance of an action by an agent. We will let $\mathsf{Do}(A, Act)$ be a description of the event consisting of the agent denoted by A performing the action denoted by Act.[5] (We will assume that the set of possible agents is the same as the set of possible knowers.) We will want $\mathsf{Do}(A, Act)$ to be the standard way of referring to the event of A's carrying out the action Act, so Do will be a rigid function. Hence, $\mathsf{Do}(A, Act)$ will be a

[4]This amounts to an assumption that all events are deterministic, which might seem to be an unnecessary limitation. From a pragmatic standpoint, however, it doesn't matter whether we say that a given event is nondeterministic, or we say that it is deterministic but no one knows precisely what the outcome will be. If we treated events as being nondeterministic, we could say that an agent knows exactly what situation he is in, but, because $:E$ is nondeterministic, he doesn't know what situation would result if $:E$ occurs. It would be completely equivalent, however, to say that $:E$ is deterministic, and that the agent *does not* know exactly what situation he is in because he doesn't know what the result of $:E$ would be in that situation.

[5]It would be more precise to say that $\mathsf{Do}(A, Act)$ names a *type* of event rather than an individual event, since an agent can perform the same action on different occasions. We would then say that Res and R apply to event types. We will let the present usage stand, however, since we have no need to distinguish event types from individual events.

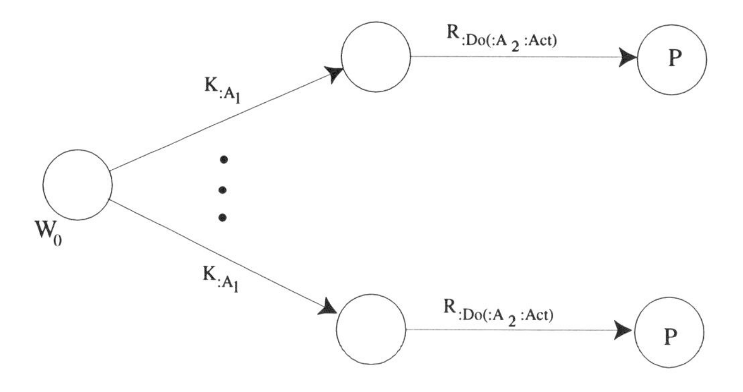

Figure 5. $\mathsf{True}(\mathsf{Know}(A_1,\mathsf{Res}(\mathsf{Do}(A_2,Act),P))) \equiv$
$\forall w_1(K(:A_1,W_0,w_1) \supset$
$\exists w_2(R(:\mathsf{Do}(:A_2,:Act),w_1,w_2) \wedge T(w_2,P)))$

rigid designator of an event if A is a rigid designator of an agent and Act a rigid designator of an action.

Many actions can be thought of as general procedures applied to particular objects. Such a general procedure will be represented by a function that maps the objects to which the procedure is applied into the action of applying the procedure to those objects. For instance, if Dial represents the general procedure of dialing combinations of safes, S a safe, and Comb(S) the combination of S, then Dial(Comb(S), S) represents the action of dialing the combination Comb(S) on the safe S, and D(A, Dial(Comb(S), S)) represents the event of A's dialing the combination Comb(S) on the safe S.

This formalism gives us the ability describe an agent's knowledge of the effects of carrying out an action. In the object language, we can express the claim that A_1 knows that P would result from A_2's doing Act by saying that $\mathsf{Know}(A_1, \mathsf{Res}(\mathsf{Do}(A_2, Act), P))$ is true. The possible-world analysis of this statement is that, for every world compatible with what A_1 knows in the actual world, there is a world that is the result of A_2's doing Act and in which P is true (see Figure 5). Formally, this is expressed by

$$\forall w_1(K(:A_1, W_0, w_1) \supset \exists w_2(R(:\mathsf{Do}(:A_2,:Act), w_1, w_2) \wedge T(w_2, P))),$$

if we assume that A_1, A_2, and Act are rigid designators.

In addition to simple, one-step actions, we will want to talk about complex combinations of actions. We will therefore introduce expressions into the object language for action sequences, conditionals, and iteration. If P is a formula, and Act_1 and Act_2 are action descriptions, then (Act_1;Act_2), If(P,Act_1,Act_2), and While(P,Act_1) will also be action descriptions. Roughly speaking, (Act_1;Act_2) describes the sequence of actions consisting of Act_1 followed by Act_2. If(P,Act_1,Act_2) describes the conditional action of doing Act_1 if P is true, otherwise doing Act_2. While(P,Act_1) describes the iterative action of repeating Act_1 as long as P is true.

Defining denotations for these complex action descriptions is somewhat problematical. The difficulty comes from the fact that, whenever we have an action described as a sequence of subactions, any expression used in specifying one of the subactions needs to be interpreted relative to the situation in which that subaction is carried out. For instance, if Puton(X, Y) denotes the action of putting X on Y, Stack denotes a stack of blocks, Table denotes a table, and Top picks out the top block of a stack, we would want the execution of

(Puton(Top(Stack), Table); Puton(Top(Stack), Table)

to result in what were initially the top two blocks of the stack being put on the table, rather than what was initially the top block being put on the table twice. The second occurrence of Top(Stack) should be interpreted with respect to the situation in which the first block has already been removed. The problem is that, in general, what situation exists after one step of a sequence of actions has been excecuted depends on who the agent is. If John picks up a certain block, *he* will be holding the block; if, however, Mary performs the same action, *she* will be holding the block. If an action description refers to "the block Mary is holding," exactly which block it is may depend on which agent is carrying out the action, but this is not specified by the action description.

One way of getting around these difficulties conceptually would be to treat actions as functions from agents to events, but notational problems would remain nevertheless. We will therefore choose a different solution: treating complex actions as "virtual individuals" (Scott 1970), or pseudoentities. That is, complex action descriptions will not be treated as referring expressions in themselves, but only as component parts of more complex referring expressions. In particular, if Act is a complex action description (and A denotes an agent), we will treat the event description Do(A, Act), but not Act itself, as having a denotation. Complex action descriptions will be permitted to occur only as part of such event descriptions, and we will define the denotations of the event descriptions in a way that eliminates reference to complex actions. We will, however, continue to treat actions as real entities that can be quantified over, and simple action descriptions such as $\mathsf{Dial}(\mathsf{Comb}(S), S)$ will still be considered to denote actions.

The denotations of event descriptions formed from conditional and iterative action descriptions can be defined as follows in terms of the denotations of event descriptions formed from action sequence descriptions:

R3. $\forall w_1, t_1, t_2, t_3, p_1$
$$((T(w_1, p_1) \supset (D(w_1, \mathsf{Do}(t_1, \mathsf{If}(p_1, t_2, t_3))) = D(w_1, \mathsf{Do}(t_1, t_2)))) \wedge (\neg T(w_1, p_1) \supset (D(w_1, \mathsf{Do}(t_1, \mathsf{If}(p_1, t_2, t_3))) = D(w_1, \mathsf{Do}(t_1, t_3)))))$$

R4. $\forall w_1, t_1, t_2, p_1$
$$(D(w_1, \mathsf{Do}(t_1, \mathsf{While}(p_1, t_2))) = D(w_1, \mathsf{Do}(t_1, \mathsf{If}(p_1, (t_2; \mathsf{While}(p_1, t_2)), \mathsf{Nil}))))$$

R3 says that performing the conditional action $\mathsf{If}(P, Act_1, Act_2)$ results in the same event as carrying out Act_1 in a situation where P is true or carrying out Act_2 in a situation where P is false. R4 says that performing $\mathsf{While}(P, Act)$ always results in the same event as

$$\mathsf{If}(\mathrm{P}, (Act; \mathsf{While}(\mathrm{P}, Act)), \mathsf{Nil})$$

where Nil denotes the null action. In other words, doing $\mathsf{While}(P, Act)$ is equivalent to doing Act followed by $\mathsf{While}(P, Act)$ if P is true, otherwise doing nothing—i.e., doing Act as long as P remains true.

To define the denotation of events that consist of carrying out action sequences, we need some notation for talking about sequences of events. First, we will let ";" be a polymorphic operator in the object language, creating descriptions of event sequences in addition to action sequences. Speaking informally, if E_1 and E_2 are event descriptions, then $(E_1; E_2)$ names the event sequence consisting of E_1 followed by E_2, just as $(Act_1;Act_2)$ names the action sequence consisting of Act_1 followed by Act_2. In the metalanguage, event sequences will be indicated with angle brackets, so that $\langle :E_1, :E_2 \rangle$ will mean $:E_1$ followed by $:E_2$. The denotations of expressions involving action and event sequences are then defined by the following axioms:

R5. $\forall w_1, t_1, t_2, t_3$
$$(D(w_1, \mathsf{Do}(t_1, (t_2; t_3))) = D(w_1, (\mathsf{Do}(t_1, t_2); \mathsf{Do}(@(D(w_1, t_1)), t_3))))$$

R6. $\forall w_1, w_2, t_1, t_2$
$$(R(D(w_1, t_1), w_1, w_2) \supset$$

$$(D(w_1,(t_1;t_2)) = \langle D(w_1,t_1), D(w_2,t_2)\rangle))$$

R5 says that the event consisting of an agent A's performance of the action sequence Act_1 followed by Act_2 is simply the event sequence that consists of A's carrying out Act_1 followed by his carrying out Act_2. Note that, in the description of the second event, the agent is picked out by the expression $@(D(w_1, A))$, which guarantees that we get the same agent as in the first event, in case the original term picking out the agent changes its denotation after the first event has happened. R6 then defines the denotation of an event sequence description $(E_1; E_2)$ as the sequence comprising the denotation of E_1 in the original situation followed by the denotation of E_2 in the situation resulting from the occurrence of E_1. If there is no situation that results from the occurence of E_1, we leave the denotation of $(E_1; E_2)$ undefined.

Finally, we need to define the accessibility relation R for event sequences and for events in which the null action is carried out.

R7. $\forall w_1, w_2, e_1, e_2$
$$(R(\langle e_1, e_2\rangle, w_1, w_2) \equiv \exists w_3(R(e_1, w_1, w_3) \wedge R(e_2, w_3, w_2)))$$

R8. $\forall w_1, a_1(R(:\mathsf{Do}(a_1, :\mathsf{Nil}), w_1, w_1))$

R7 says that a situation W_2 is the result of the event sequence $\langle E_1, E_2\rangle$ occurring in W_1 if and only if there is a situation W_3 such that W_3 is the result of E_1 occurring in W_1, and W_2 is the result of E_2 occurring in W_3.[6] We will regard Nil as a rigid designator in the object language for the null action, so :Nil will be its metalanguage counterpart. R8, therefore, says that in any situation the result of doing nothing is the same situation.

3.5 An Integrated Theory of Knowledge and Action

The Dependence of Action on Knowledge

As we pointed out in the introduction, knowledge and action interact in two principal ways: (1) knowledge is often required prior to taking action; (2) actions can change what is known. In regard to the first, we need to consider knowledge prerequisites as well as physical prerequisites for actions. Our main thesis is that the knowledge prerequisites for an action can be analyzed as a matter of knowing what action to

[6] R7 guarantees that the sequences $\langle\langle E_1, E_2\rangle, E_3\rangle$ and $\langle E_1, \langle E_2, E_3\rangle\rangle$ always define the same accessibility relation on situations; so, just as one would expect, we can regard sequence operators as being associative. Thus, when we have a sequence of more than two events or actions, we will not feel obliged to indicate a pairwise grouping.

take. Recall the example of trying to open a locked safe. Why is it that, for an agent to achieve this goal by using the plan "Dial the combination of the safe," he must know the combination? The reason is that an agent could know that dialing the combination of the safe would result in the safe's being open, but still not know what to do because he does not know what the combination of the safe is. A similar analysis applies to knowing a telephone number in order to call someone on the telephone or knowing a password in order to gain access to a computer system.

It is important to realize that even mundane actions that are not usually thought of as requiring any special knowledge are no different from the examples just cited. For instance, none of the AI problem-solving systems that have dealt with the blocks world have tried to take into account whether the robot possesses sufficient knowledge to be able to move block A to point B. Yet, if a command were phrased as "Move my favorite block back to its original position," the system could be just as much in the dark as with "Dial the combination of the safe." If the system does not know what actions satisfy the description, it will not be able to carry out the command. The only reason that the question of knowledge seems more pertinent in the case of dialing combinations and telephone numbers is that, in the contexts in which these actions naturally arise, there is usually no presumption that the agent knows what action fits the description. An important consequence of this view is that the specification of an action will normally not need to include anything about knowledge prerequisites. These will be supplied by a general theory of using actions to achieve goals. What we will need to specify are the conditions under which an agent knows what action is referred to by an action description.

In our possible-world semantics for knowledge, the usual way of knowing what entity is referred to by a description B is by having some description C that is a rigid designator, and by knowing that $B = C$. (Note, that if B itself is a rigid designator, it can be used for C.) In particular, knowing what action is referred to by an action description means having a rigid designator for the action described. But, if this is all the knowledge that is required for carrying out the action, then a rigid designator for an action must be an *executable description* of the action—in the same sense that a computer program is an executable description of a computation to an interpreter for the language in which the program is written.

Often the actions we want to talk about are mundane general procedures that we would be willing to assume everyone knows how to perform. Dialing a telephone number or the combination of a safe is a typical example. In many of these cases, if an agent knows the general procedure and what objects the procedure is to be applied to, then he knows everything that is relevant to the task. In such cases, the function that represents the general procedure will be a rigid function, so that, if the arguments of the function are rigid designators, the term consisting of the function applied to the arguments will be a rigid designator. Hence, knowing what objects the arguments denote will amount to knowing what action the term refers to. We will treat dialing the combination of a safe, or dialing a telephone number as being this type of procedure. That is, we assume that anyone who knows what combination he is to dial and what safe he is to dial it on thereby knows what action he is to perform.

There are other procedures we might also wish to assume that anyone could perform, but that cannot be represented as rigid functions. Suppose that, in the blocks world, we let $\mathsf{Puton}(B, C)$ denote the action of putting B on C. Even though we would not want to question anyone's ability to perform Puton in general, knowing what objects B and C are will not be sufficient to perform $\mathsf{Puton}(B, C)$; knowing *where* they are is also necessary. We could have a special axiom stating that knowing what action $\mathsf{Puton}(B, C)$ is requires knowing where B and C are, but this will be superfluous if we simply assume that everyone knows the definition of Puton in terms of more primitive actions. If we define $\mathsf{Puton}(X, Y)$ as something like

$\quad$ (Movehand(Location(X));
$\quad$ Grasp;
$\quad$ Movehand(Location(Top(Y)));
$\quad$ Ungrasp),

then we can treat Movehand, Grasp, and Ungrasp as rigid functions, and we can see that executing Puton requires knowing where the two objects are because their locations are mentioned in the definition. So, although Puton itself is not a rigid function, we can avoid having a special axiom stating what the knowledge prerequisites of Puton are by defining Puton as a sequence of actions represented by rigid functions.

To formalize this theory, we will introduce a new object language operator Can. $\mathsf{Can}(A, Act, P)$ will mean that A can achieve P by performing Act, in the sense that A knows how to achieve P by performing Act. We will not give a possible-world semantics for Can directly; instead we will give a definition of Can in terms of Know and Res, which we can use in reasoning about Can to transform a problem into terms of possible worlds.

In the simplest case, an agent A can achieve P by performing Act

if he knows what action *Act* is, and he knows that P would be true as a result of his performing *Act*. In the object language, we can express this fact by

$$\forall a(\exists x(\mathsf{Know}(a,((x = Act) \wedge \mathsf{Res}(\mathsf{Do}(a, Act), P)))) \supset \mathsf{Can}(a,Act,P)).$$

We cannot strengthen this assertion to a biconditional, however, because that would be too stringent a definition of Can for complex actions. It would require the agent to know from the very beginning of his action exactly what he is going to do at every step. In carrying out a complex action, though, an agent may take some initial action that results in his acquiring knowledge about what to do later.

For an agent to be able to achieve a goal by performing a complex action, all that is really neccessary is that he know what to do first, and that he know that he will know what to do at each subsequent step. So, for any action descriptions *Act* and Act_1, the following formula also states a condition under which an agent can achieve P by performing *Act*:

$$\begin{aligned}&\forall a(\exists x(\mathsf{Know}(a,((\mathsf{Do}(a,(x;Act_1)) = \mathsf{Do}(a, Act))\wedge\\ &\qquad\qquad \mathsf{Res}(\mathsf{Do}(a,x),\mathsf{Can}(a,Act_1,P))))) \supset\\ &\quad \mathsf{Can}(a,Act,P)).\end{aligned}$$

This says that A can achieve P by doing *Act* if there is an action X such that A knows that his execution of the sequence X followed by Act_1 would be equivalent to his doing *Act*, and thathis doing X would result in his being able to achieve P by doing Act_1.

Finally, with the following metalanguage axiom we can state that these are the only two conditions under which an agent can use a particular action to achieve a goal:

$$\begin{aligned}&\text{C1. } \forall w_1,t_1,t_2,t_3,p_1\\ &\quad ((t_2 = @(D(w_1,t_1))) \supset\\ &\quad (T(w_1,\mathsf{Can}(t_1,t_3,p_1)) \equiv\\ &\quad (T(w_1,\mathsf{Exist}(\mathsf{X},\mathsf{Know}(t_1,\mathsf{And}(\mathsf{Eq}(\mathsf{X},\mathsf{t}_3),\mathsf{Res}(\mathsf{Do}(\mathsf{t}_2,\mathsf{t}_3),\mathsf{p}_1)))))\vee\\ &\quad \exists t_4(T(w_1,\mathsf{Exist}(\mathsf{X},\mathsf{Know}(t_1,\mathsf{And}(\mathsf{Eq}(\mathsf{Do}(t_2,(\mathsf{X};t_4)),\mathsf{Do}(t_2,t_3)),\\ &\qquad\qquad \mathsf{Res}(\mathsf{Do}(t_2,\mathsf{X}),\\ &\qquad\qquad\quad \mathsf{Can}(t_2,t_4,p_1)))))))))))\end{aligned}$$

Letting $t_1 = A$, $t_2 = A_1$, and $t_3 = Act$, C1 says that, for any formula P, if A_1 is the standard identifier of the agent denoted by A, then A can achieve P by doing *Act* if and only if one of the following conditions is met: (1) A knows what action *Act* is and knows that P would be true as a result of A_1's (i.e., his) doing *Act*, or (2) there is an action description $t_4 = Act_1$ such that, for some action X, A knows that A_1's doing X followed by Act_1 is the same event as his doing *Act* and knows that A_1's doing X would result his being able to achieve P by doing Act_1.

As a simple illustration of these concepts, we will show how to derive the fact that an agent can open a safe, given the premise that he knows the combination. To do this, the only additional fact we need is that, if an agent does dial the correct combination of a safe, the safe will then be open:

$$\text{D1. } \forall w_1, a_1, x_1 \\ (\text{:Safe}(x_1) \supset \\ \exists w_2(R(\text{:Do}(a_1, \text{:Dial}(\text{:Comb}(w_1, x_1), x_1)), w_1, w_2) \wedge \\ \text{:Open}(w_2, x_1)))$$

D1 says that, for any possible world W_1, any agent $:A$, and any safe $:S$, there is a world W_2 that is the result of $:A$'s dialing the combination of $:S$ on $:S$ in W_1, and in which $:S$ is open. The important point about this axiom, is that the function :Comb (which picks out the combination to a safe) depends on what possible world it is evaluated in, while :Dial (the function that maps a combination and a safe into the action of dialing the combination on the safe) does not. Thus we are implicitly assuming that, given a particular safe, there may be some doubt as to what its combination is, but, given a combination and a safe, there exists no possible doubt as to what action dialing the combination on the safe is. (We also simplify matters by omitting the possible world-argument to :Safe, so as to avoid raising the question of knowing whether something is a safe.) Since this axiom is asserted to hold for all possible worlds, we are in effect assuming that it is common knowledge.

Now we show that, for any safe, if the agent A knows its combination, he can open the safe by dialing that combination; or, more precisely, for all X, if X is a safe and there is some Y, such that A knows that Y is the combination of X, then A can open X by dialing the combination of X on X:

Prove: True(All(X, Imp(And(Safe(X), Exist(Y, Know(A, Eq(Y, Comb(X))))))
Can(A, Dial(Comb(X), X), Open(X))))

1.	$T(W_0,$ And(Safe(@(x_1)), Exist(Y, Know(A, Eq(Y, Comb(@(x_1))))))))	ASS
2.	$\text{:Safe}(x_1)$	1,L2,L9
3.	$\forall w_1(K(:A(W_0), W_0, w_1) \supset (:C = \text{:Comb}(w_1, x_1)))$	1,L2,L7,K1,L11, L13,L10,L12
4.	$K(:A(W_0), W_0, w_1)$	ASS
5.	$:C = \text{:Comb}(w_1, x_1)$	3,4

	Formula	Justification
6.	$\mathsf{:Dial}(:C, x_1) = \mathsf{:Dial}(\mathsf{:Comb}(w_1, x_1), x_1)$	5
7.	$T(w_1,$ $\quad \mathsf{Eq}(@(\mathsf{:Dial}(:C, x_1)),$ $\quad\quad \mathsf{Dial}(\mathsf{Comb}(@(x_1)), @(x_1))))$	L10,L12,L12a, L13
8.	$\exists w_2(R(\mathsf{:Do}(:A(W_0),$ $\quad\quad\quad \mathsf{:Dial}(\mathsf{:Comb}(w_1, x_1), x_1)),$ $\quad\quad w_1, w_2) \wedge$ $\quad \mathsf{:Open}(w_2, x_1)))$	2,D1
9.	$T(w_1,$ $\quad \mathsf{Res}(\mathsf{Do}(@(D(W_0, \mathsf{A})),$ $\quad\quad\quad \mathsf{Dial}(\mathsf{Comb}(@(x_1)), @(x_1))),$ $\quad\quad \mathsf{Open}(@(x_1))))$	L11,L10,L12a, L9, R2
10.	$T(w_1,$ $\quad \mathsf{And}(\mathsf{Eq}(@(\mathsf{:Dial}(:C, x_1)),$ $\quad\quad\quad \mathsf{Dial}(\mathsf{Comb}(@(x_1)), @(x_1))),$ $\quad\quad \mathsf{Res}(\mathsf{Do}(@(D(W_0, \mathsf{A})),$ $\quad\quad\quad\quad \mathsf{Dial}(\mathsf{Comb}(@(x_1)), @(x_1))),$ $\quad\quad\quad \mathsf{Open}(@(x_1)))))$	7,9,L2
11.	$K(:A(W_0), W_0, w_1) \supset$ $T(w_1,$ $\quad \mathsf{And}(\mathsf{Eq}(@(\mathsf{:Dial}(:C, x_1)),$ $\quad\quad\quad \mathsf{Dial}(\mathsf{Comb}(@(x_1)), @(x_1))),$ $\quad\quad \mathsf{Res}(\mathsf{Do}(@(D(W_0, \mathsf{A})),$ $\quad\quad\quad\quad \mathsf{Dial}(\mathsf{Comb}(@(x_1)), @(x_1))),$ $\quad\quad\quad \mathsf{Open}(@(x_1)))))$	DIS(4,10)
12.	$T(W_0,$ $\quad \mathsf{Know}(A,$ $\quad\quad \mathsf{And}(\mathsf{Eq}(@(\mathsf{:Dial}(:C, x_1)),$ $\quad\quad\quad\quad \mathsf{Dial}(\mathsf{Comb}(@(x_1)), @(x_1))),$ $\quad\quad\quad \mathsf{Res}(\mathsf{Do}(@(D(W_0, \mathsf{A})),$ $\quad\quad\quad\quad\quad \mathsf{Dial}(\mathsf{Comb}(@(x_1)), @(x_1))),$ $\quad\quad\quad\quad \mathsf{Open}(@(x_1))))))$	11,L11,K1
13.	$T(W_0,$ $\quad \mathsf{Exist}(\mathsf{X},$ $\quad\quad \mathsf{Know}(\mathsf{A},$ $\quad\quad\quad \mathsf{And}(\mathsf{Eq}(\mathsf{X},$ $\quad\quad\quad\quad\quad \mathsf{Dial}(\mathsf{Comb}(@(x_1)),$ $\quad\quad\quad\quad\quad\quad @(x_1))),$ $\quad\quad\quad\quad \mathsf{Res}(\mathsf{Do}(@(D(W_0, \mathsf{A})),$ $\quad\quad\quad\quad\quad\quad \mathsf{Dial}(\mathsf{Comb}(@(x_1)),$ $\quad\quad\quad\quad\quad\quad\quad @(x_1))),$	12,L7

```
                    Open(@(x1)))))
14. T(W0,                                        13,C1
      Can(A,
            Dial(Comb(@(x1)), @(x1)),
            Open(@(x1))))
15. T(W0,                                        DIS(1,14)
      And(Safe(@(x1),
            Exist(Y, Know(A, Eq(Y, Comb(@(x1)))))))) ⊃
    T(W0,
      Can(A, Dial(Comb(@(x1)), @(x1)), Open(@(x1))))
16. True(All(X,                                  15,L4,L8,L1
            Imp(And(Safe(X),
                    Exist(Y,
                          Know(A,
                                Eq(Y, Comb(X))))))
                Can(A, Dial(Comb(X), X), Open(X))))
```

Suppose that x_1 is a safe and there is some C that A knows to be the combination of x_1 (Lines 1–3). Suppose w_1 is a world that is compatible with what A knows in the actual world, W_0 (Line 4). Then C is the combination of x_1 in w_1 (Line 5), so dialing C on x_1 is the same action as dialing the combination of x_1 on x_1 in w_1 (Lines 6 and 7). By axiom D1, A's dialing the combination of x_1 on x_1 in w_1 will result in x_1's being open (Lines 8 and 9). Since w_1 was an arbitrarily chosen world compatible with what A knows in W_0, it follows that in W_0 A knows dialing C on x_1 to be the act of dialing the combination of x_1 on x_1 and that his dialing the combination of x_1 on x_1 will result in x_1's being open (Lines 10–12). Hence, A knows what action dialing the combination of x_1 on x_1 is, and that his dialing the combination of x_1 on x_1 will result in x_1's being open (Line 13). Therefore A can open x_1 by dialing the combination of x_1 on x_1, provided that x_1 is a safe and he knows the combination of x_1 (Lines 14 and 15). Finally, since x_1 was chosen arbitrarily, we conclude that A can open any safe by dialing the combination, provided he knows the combination (Line 16).

The Effects of Action on Knowledge

In describing the effects of an action on what an agent knows, we will distinguish actions that give the agent new information from those that do not. Actions that provide an agent with new information will be called *informative* actions. An action is informative if an agent would know more about the situation resulting from his performing the action after performing it than before performing it. In the blocks world, looking inside a box could be an informative action, but moving a block would probably not, because an agent would normally know no more after moving the block than he would before moving it. In the real world there are probably no actions that are never informative, because all physical processes are subject to variation and error. Nevertheless, it seems clear that we do and should treat many actions as noninformative from the standpoint of planning.

Even if an action is not informative in the sense we have just defined, performing the action will still alter the agent's state of knowledge. If the agent is aware of his action, he will know that it has been performed. As a result, the tense and modality of many of the things he knows will change. For example, if before performing the action he knows that P is true, then after performing the action he will know that P was true before he performed the action. Similarly, if before performing the action he knows that P would be true after performing the action, then afterwards he will know that P is true.

We can represent this very elegantly in terms of possible worlds. Suppose $:A$ is an agent and $:E_1$ an event that consists in $:A$'s performing some noninformative action. For any possible worlds W_1 and W_2 such that W_2 is the result of $:E_1$'s happening in W_1, the worlds that are compatible with what $:A$ knows in W_2 are exactly the worlds that are the result of $:E_1$'s happening in some world that is compatible with what $:A$ knows in W_1. In formal terms, this is

$$\begin{aligned}&\forall w_1, w_2(R(:E, w_1, w_2) \supset \\ &\qquad \forall w_3(K(:A, w_2, w_3) \equiv \\ &\qquad\qquad \exists w_4(K(:A, w_1, w_4) \land R(:E, w_4, w_3)))),\end{aligned}$$

which tells us exactly how what $:A$ knows after $:E_1$ happens is related to what $:A$ knows before $:E_1$ happens.

We can try to get some insight into this analysis by studying Figure 6. Sequences of possible situations connected by events can be thought of as possible courses of events. If W_1 is an actual situation in which $:E_1$ occurs, thereby producing W_2, then W_1 and W_2 comprise a subsequence of the actual course of events. Now we can ask what other courses of events are compatible with what $:A$ knows in W_1 and in W_2. Suppose that W_4 and W_3 are connected by $:E_1$ in a course of events that is compatible with what $:A$ knows in W_1. Since $:E_1$ is not informative for $:A$, the only sense in which his knowledge is increased by $:E_1$ is that he knows that $:E_1$ has occurred. Since $:E_1$ occurs at the corresponding place in the course of events that includes W_4 and W_3, this course of events will still be compatible with everything $:A$ knows

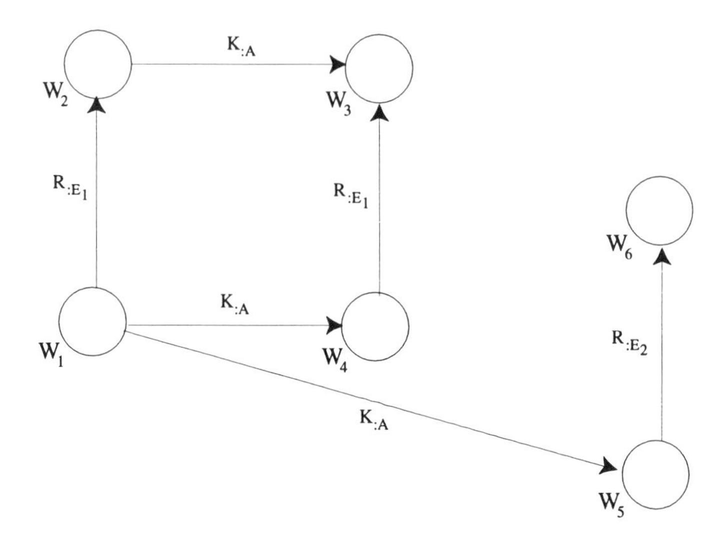

Figure 6. The effect of a noninformative action on the agent's knowledge

in W_2. However, the appropriate "tense shift" takes place. In W_1, W_4 is a possible alternative present for $:A$, and W_3 is a possible alternative future. In W_2, W_3 is a possible alternative present for $:A$, and W_4 is a possible alternative past.

Next consider a different course of events that includes W_5 and W_6 connected by a different event, $:E_2$. This course of events might be compatible with what $:A$ knows in W_1 if he is not certain what he will do next, but, after $:E_1$ has happened and he knows that it has happened, this course of events is no longer compatible with what he knows. Thus, W_6 is not compatible with what $:A$ knows in W_2. We can see, then, that even actions that provide the agent with no new information from the outside world still filter out for him those courses of events in which he would have performed actions other than those he actually did.

The idea of a filter on possible courses of events also provides a good picture of informative actions. With these actions, though, the filter is even stronger, since they not only filter out courses of events that differ from the actual course of events as to what event has just occurred, but they also filter out courses of events that are incompatible with the information furnished by the action. Suppose $:E$ is an event that consists in $:A$'s performing an informative action, such that the information gained by the agent is whether the formula P is true. For

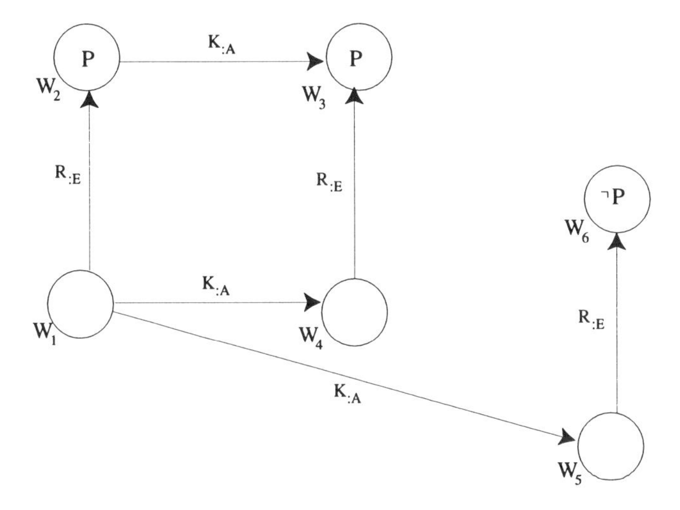

Figure 7. The effect of an informative action on the agent's knowledge

any possible worlds W_1 and W_2 such that W_2 is the result of $:E$'s happening in W_1, the worlds that are compatible with what $:A$ knows in W_2 are exactly those worlds that are the result of $:E$'s happening in some world that is compatible with what $:A$ knows in W_1, *and in which P has the same truth-value as in* W_2:

$$\begin{aligned}&\forall w_1, w_2(R(:E, w_1, w_2) \supset \\ &\qquad \forall w_3(K(:A, w_2, w_3) \equiv \\ &\qquad\qquad (\exists w_4(K(:A, w_1, w_4) \wedge R(:E, w_4, w_3)) \wedge \\ &\qquad\qquad (T(w_2, P) \equiv T(w_3, P)))))\end{aligned}$$

It is this final condition that distinguishes informative actions from those that are not.

Figure 7 illustrates this analysis. Suppose W_1 and W_2 are connected by $:E$ and are part of the actual course of events. Suppose, further, that P is true in W_2. Let W_4 and W_3 also be connected by $:E$, and let them be part of a course of events that is compatible with what $:A$ knows in W_1. If P is true in W_3 and the only thing $:A$ learns about the world from $:E$ (other than that it has occurred) is whether P is true, this course of events will then still be compatible with what $:A$ knows after $:E$ has occurred. That is, W_3 will be compatible with what $:A$ knows in W_2. Suppose, on the other hand, that W_5 and W_6 form part of a similar course of events, except that P is false in W_6. If $:A$ does not know in W_1 whether P would be true after the occurrence of $:E$, this course of events will also be compatible with what he knows in W_1. After $:E$ has occurred, however, he will know that P is true consequently, this course of events will no longer be compatible with what he knows. That is, W_6 will not be compatible with what $:A$ knows in W_2.

It is an advantage of this approach to describing how an action affects what an agent knows that not only do we specify what he learns from the action, but also what he does not learn. Our analysis gives us necessary, as well as sufficient, conditions for $:A$'s knowing that P is true after event $:E$. In the case of an action that is not informative, we can infer that, unless $:A$ knows before performing the action whether P would be true, he will not know afterwards either. In the case of an informative action such that what is learned is whether Q is true, he will not know whether P is true unless he does already—or knows of some dependence of P on Q.

Within the context of this possible-world analysis of the effects of action on knowledge, we can formalize the requirements for a test that we presented in Section 3.1. Suppose that Test is the action of testing the acidity of a particular solution with blue litmus paper, Red is a propositional constant (a predicate of zero arguments) whose truth depends on the color of the litmus paper, and Acid is a propositional constant whose truth depends on whether the solution is acidic. The relevent fact about Test is that the paper will be red after an agent A performs the test if and only if the solution is acidic at the time the test is performed:

$$(\mathsf{Acid} \supset \mathsf{Res}(\mathsf{Do}(A, \mathsf{Test}), \mathsf{Red})) \wedge (\neg\mathsf{Acid} \supset \mathsf{Res}(\mathsf{Do}(A, \mathsf{Test}), \neg\mathsf{Red}))$$

In Section 3.1 we listed three conditions that ought to be sufficient for an agent to determine, by observing the outcome of a test, whether some unobservable precondition holds; in this case, for A to determine whether Acid is true by observing whether Red is true after Test is performed:

(1) After A performs Test, he knows whether Red is true.
(2) After A performs Test, he knows that he has just performed Test.
(3) A knows that Red will be true after Test is performed just in case Acid was true before it was performed.

Conditions (1) and (2) will be satisfied if Test is an informative action, such that the knowledge provided is whether Red is true in the resulting situation:

$$\begin{array}{ll}\text{T1.} & \forall w_1, w_2, a_1 \\ & \quad (R(\mathsf{Do}(a_1, :\mathsf{Test}), w_1, w_2) \supset \\ & \quad\; \forall w_3(K(a_1, w_2, w_3) \equiv \\ & \qquad\quad (\exists w_4(K(a_1, w_1, w_4) \wedge R(:\mathsf{Do}(a_1, :\mathsf{Test}), w_4, w_3)) \wedge\end{array}$$

$$(:\mathsf{Red}(w_2) \equiv :\mathsf{Red}(w_3)))))$$

If :Red and :Test are the metalanguage analogues of Red and Test, T1 says that for any possible worlds W_1 and W_2 such that W_2 is the result of an agent's performing Test in W_1, the worlds that are compatible with what the agent knows in W_2 are exactly those that are the result of his performing Test in some world that is compatible with what he knows in W_1, and in which Red has the same truth-value as in W_2. In other words, after performing Test, the agent knows that he has done so and he knows whether Red is true in the resulting situation. As with our other axioms, the fact that it holds for all possible worlds makes it common knowledge.

Thus, A can use Test to determine whether the solution is acid, provided that (3) is also satisfied. We can state this very succinctly if we make the further assumption that A knows that performing the test does not affect the acidity of the solution.[7] Given the axiom T1 for test, it is possible to show that

$$\mathsf{Acid} \supset \mathsf{Res}(\mathsf{Do}(A, \mathsf{Test}), \mathsf{Know}(A, \mathsf{Acid}))$$

and

$$\neg\mathsf{Acid} \supset \mathsf{Res}(\mathsf{Do}(A, \mathsf{Test}), \mathsf{Know}(A, \neg\mathsf{Acid}))$$

are true, provided that

$$\mathsf{Know}(A, (\mathsf{Acid} \supset \mathsf{Res}(\mathsf{Do}(A, \mathsf{Test}), (\mathsf{Acid} \land \mathsf{Red}))))$$

and

$$\mathsf{Know}(A, (\neg\mathsf{Acid} \supset \mathsf{Res}(\mathsf{Do}(A, \mathsf{Test}), (\neg\mathsf{Acid} \land \neg\mathsf{Red}))))$$

are both true and A is a rigid designator. We will carry out the proof in one direction, showing that, if the solution is acidic, after the test has been conducted the agent will know that it is acidic.

Given: True(Know(A, Imp(Acid, Res(Do(A, Test), And(Acid, Red)))))
True(Know(A, Imp(Not(Acid), Res(Do(A, Test),
And(Not(Acid), Not(Red))))))
True(Acid)

Prove: True(Res(Do(A, Test), Know(A, Acid)))

[7] We have to add this extra condition to be able to infer that the agent knows whether the solution *is* acidic, instead of merely that he knows whether it *was* acidic. The latter is a more general characteristic of tests, since it covers destructive as well as nondestructive tests. We have not, however, introduced any temporal operators into the object language that would allow us to make such a statement, although there would be no difficulty in stating the relevant conditions in the object language. Indeed, this is precisely what is done by axioms such as T1.

1.	$\forall w_1(K(:A, W_0, w_1) \supset (:\mathsf{Acid}(w_1) \supset \exists w_2(R(:\mathsf{Do}(:A, :\mathsf{Test}), w_1, w_2) \wedge :\mathsf{Acid}(w_2) \wedge :\mathsf{Red}(w_2))))$	Given,L1,L4,R2, L2,L9,L12,L11a
2.	$\forall w_1(K(:A, W_0, w_1) \supset (\neg:\mathsf{Acid}(w_1) \supset \exists w_2(R(:\mathsf{Do}(:A, :\mathsf{Test}), w_1, w_2) \wedge \neg:\mathsf{Acid}(w_2) \wedge \neg:\mathsf{Red}(w_2))))$	Given,L1,L4,R2,L2, L6,L9,L12,L11a
3.	$:\mathsf{Acid}(W_0)$	L1,L9
4.	$:\mathsf{Acid}(W_0) \supset \exists w_2(R(:\mathsf{Do}(:A, :\mathsf{Test}), W_0, w_2) \wedge :\mathsf{Acid}(w_2) \wedge :\mathsf{Red}(w_2))$	1,K2
5.	$R(:\mathsf{Do}(:A, :\mathsf{Test}), W_0, W_1)$	3,4
6.	$:\mathsf{Red}(W_1)$	3,4
7.	$\forall w_2(K(:A, W_1, w_2) \equiv (\exists w_3(K(:A, W_0, w_3) \wedge R(:\mathsf{Do}(:A, :\mathsf{Test}), w_3, w_2)) \wedge (:\mathsf{Red}(W_1) \equiv :\mathsf{Red}(w_2))))$	5,T1
8.	$K(:A, W_1, w_2)$	ASS
9.	$K(:A, W_0, W_3)$	7,8
10.	$R(:\mathsf{Do}(:A, :\mathsf{Test}), W_3, w_2)$	7,8
11.	$:\mathsf{Red}(W_1) \equiv :\mathsf{Red}(w_2)$	7,8
12.	$:\mathsf{Red}(w_2)$	6,11
13.	$\neg:\mathsf{Acid}(W_3) \supset \exists w_4(R(:\mathsf{Do}(:A, :\mathsf{Test}), W_3, w_4) \wedge \neg:\mathsf{Acid}(w_4) \wedge \neg:\mathsf{Red}(w_4))$	2,9
14.	$\neg:\mathsf{Acid}(W_3)$	ASS
15.	$R(:\mathsf{Do}(:A, :\mathsf{Test}), W_3, W_4)$	13,14
16.	$\neg:\mathsf{Red}(W_4)$	13,14
17.	$w_2 = W_4$	15,R1
18.	$\neg:\mathsf{Red}(w_2)$	16,17
19.	False	12,18
20.	$:\mathsf{Acid}(W_3)$	DIS(14,19)
21.	$:\mathsf{Acid}(W_3) \supset \exists w_4(R(:\mathsf{Do}(:A, :\mathsf{Test}), W_3, w_4) \wedge :\mathsf{Acid}(w_4) \wedge :\mathsf{Red}(w_4))$	1,9
22.	$R(:\mathsf{Do}(:A, :\mathsf{Test}), W_3, W_4)$	20,21
23.	$:\mathsf{Acid}(W_4)2$	20,21
24.	$w_2 = W_4$	15,22
25.	$:\mathsf{Acid}(w_2)$	23,24
26.	$K(:A, W_1, w_2) \supset :\mathsf{Acid}(w_2)$	DIS(8,25)

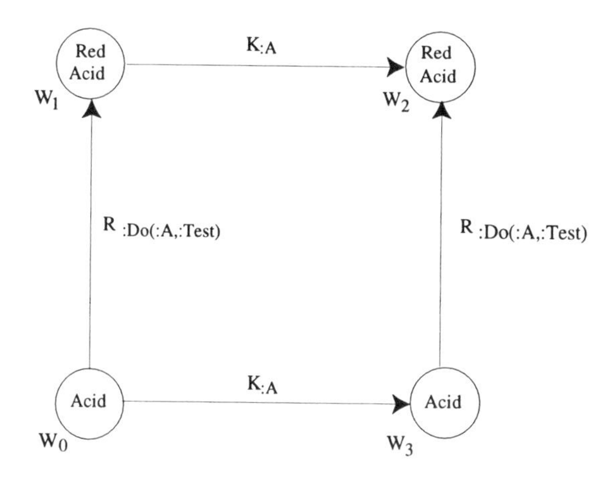

Figure 8. The effect of a test on the agent's knowledge

27.	$R(\text{:Do}(:A, \text{:Test}), W_0, W_1) \wedge$ $\forall w_2(K(:A, W_1, w_2) \supset \text{:Acid}(w_2))$	5,26
28.	$\text{True}(\text{Res}(\text{Do}(\text{A}, \text{Test}), \text{Know}(\text{A}, \text{Acid})))$	27,L9,L11a,L12, K2,R2,L1

The possible-world structure for this proof is depicted in Figure 8. Lines 1 and 2 translate the premises into the metalanguage. Since A knows that, if the solution is acidic, performing the test will result in the litmus paper's being red, it must be true in the actual world W_0 that, if the solution is acidic, performing the test will result in the litmus paper's being red (Line 3). Suppose that, in fact, the solution is acidic (Line 4). Then, if W_1 is the result of performing the test in W_0 (Line 5), the paper will be red in W_1 (Line 6). Furthermore, the worlds that are compatible with what A knows in W_1 are those that are the result of his performing the test in some world that is compatible with what he knows in W_1, and in which the paper is red if and only if it is red in W_1 (Line 7). Suppose that w_2 is a world that is compatible with what A knows in W_1 (Line 8). Then there is a W_3 that is compatible with what A knows in W_0 (Line 9), such that w_2 is the result of A's performing the test in W_3 (Line 10). The paper is red in w_2, if and only if it is red in W_1 (Line 11); therefore, it is red in w_2 (Line 12). Since A knows how the test works, if the solution were not acidic in W_3, it would not be acidic, and the paper would not be red, in w_2 (Line 13).

Now, suppose the solution were not acid in W_3 (Line 14). If W_4 is the result of A's performing the test in W_3 (Line 15), the paper would not be red in W_4 (Line 16). But w_2 is the result of A's performing the test in W_3 (Line 17), so the paper would not be red in w_2 (Line 18). We know this is false (Line 19), however, so the solution must be acidic in W_3 (Line 20). If the solution is acidic in W_3, it must also be acidic in the situation resulting from A's performing the test in W_3 (Lines 21–23), but this is w_2 (Line 24). Therefore, the solution is acidic in w_2 (Line 25). Hence, in W_1, A knows that the solution is acidic (Line 26), so in the situation resulting from A's performing the test in W_0, he knows that the solution is acidic (Line 27). In other words (Line 28), A's performing the test would result in his knowing that the solution is acidic.

By an exactly parallel argument, we could show that, if the solution were not acidic, A could also find that out by carrying out the test, so our analysis captures the sort of reasoning about tests that we described in Section 3.1, based on general principles that govern the interaction of knowledge and action.

4

Computational Models of Belief and the Semantics of Belief Sentences

WITH G. G. HENDRIX

4.1 Computational Theories and Computational Models

This chapter considers a number of problems in the semantics of belief sentences from the perspective of computational models of the psychology of belief. We present a semantic interpretation for belief sentences that is suggested by a computational model of belief, and show how this interpretation overcomes some of the difficulties of alternative approaches, especially those based on possible-world semantics. Finally, we argue that these difficulties arise from a mistaken attempt to identify the truth conditions of a sentence with what a competent speaker knows about the meaning of the sentence.

Over the years the psychology of belief and the semantics of belief sentences have provided a seemingly endless series of fascinating problems for linguists, psychologists, and philosophers. Despite all the attention that has been paid to these problems, however, there is little agreement on proposed solutions, or even on what form solutions should take. We believe that a great deal of light can be shed on the problems of belief by studying them from the viewpoint of computational models of the psychological processes and states associated with belief. The role of computational theories and computational models

The work reported herein was supported by the National Science Foundation under Grant No. MCS76-22004, and by the Defense Advanced Research Projects Agency under Contract N00039-79-C-0118 with the Naval Electronic Systems Command.

in the cognitive sciences always seems to be a matter of controversy. When such theories and models are discussed by non-computer scientists, they are frequently presented in a rather apologetic tone, with assurances and caveats that, of course, this is all oversimplified and things couldn't really be like this, *but*...

This may be the result of an unwarranted inference that anyone who takes a computational approach in one of these disciplines thereby endorses what is sometimes called the thesis of "mechanism" (Lucas 1961): that minds can be completely explained in terms of machines, which in contemporary discussions are usually taken to be computers. When the metaphysical doctrine of dualism was more widely held, the mechanism thesis could be rejected on the grounds that minds were nonphysical. It appears to be more fashionable to adhere to a materialistic metaphysics nowadays, but to hold that the way minds are embodied in brains is so complex as to be beyond all human understanding, or at least too complex to be represented by Turing machines or computer programs. On the basis of current knowledge, these questions appear to us to be completely open. The existing evidence may well have as little relevance to future discoveries as the arguments of the Greek philosophers about atomism have to modern atomic theory.

We wish to argue, however, that the usefulness of computational approaches in the cognitive sciences does not depend on how (or even whether) these questions are eventually answered. In elaborating this view, it will be helpful to make a distinction between computational theories and computational models. We will say that a theory of a cognitive process is a computational theory if it claims that the process is a computational process. The mechanism thesis can be viewed as the claim that *every* cognitive process is a computational process. Obviously, one can hold that certain cognitive processes are computational without claiming that all are, so having a particular computational theory of some cognitive process still leaves the mechanism thesis an open question.

The construction of computational theories is the most obvious use of computational ideas in the cognitive sciences. However, even without computational theories, computational models can be extremely useful. By a computational model of a cognitive process we mean a computational system whose behavior is similar to the behavior of the process in some interesting way. The important point is that one can make use of computational models without making any claims about the nature of the process being modeled. For example, the use of computational models in weather forecasting does not commit one to the claim that meteorological processes are computational.

What makes computational models in meteorology interesting is the fact that they can make useful predictions about the behavior of the system being modeled. In the cognitive sciences few models, computational or otherwise, have such predictive power, and we are hard pressed to think of any cases in which the predictions that are made can be considered useful. Thus at our current level of understanding, prediction of behavior does not appear to be the most productive role for computational models of cognitive processes.

What computational models do seem to be good for is clarification of conceptual problems. Many of the most vexing problems in the cognitive sciences are questions as to how any physical system could have the properties that cognitive systems apparently possess. Computational models can often supply answers to questions of this kind independently of empirical considerations regarding the way human (or other) cognitive systems actually function. The point is that conceptual arguments often proceed from general observations about some cognitive process to specific conclusions as to what the process must be like. One way of testing such an argument is to construct a computational model that satisfies the premises of the argument and then to see whether the conclusions apply to the model. When used in this way, a computational model may be best thought of not so much as a model of a process, but rather as a model (in the sense of "model theory" in formal logic) of the theory in which the argument is made. That is, a conceptual argument ought to be valid for all possible models that satisfy its premises, so it had better be valid for a particular computational model, independently of how closely that model resembles the cognitive process that is the "intended model."

In the remainder of this chapter we will try to apply computational models in this way to investigate some of the problems about belief and the semantics of belief sentences. First we will present a model of belief that seems to satisfy most of our pretheoretical notions. Then we will ask what implications it would have for the semantics of belief sentences, if human belief were analogous to our model. As will be seen below, this leads us to some conclusions quite different from those drawn by other authors.

4.2 Internal Languages

Before going into the details of what a computational model of belief might look like, we need to deal with a set of objections that have been raised to one of the basic assumptions we will make. The assumption is that beliefs are to be explained in terms of expressions in some sort of internal language that is not the language used externally—a "language of thought" to use Fodor's (1975) term. To possess a particular belief is to bear a certain computational relation to the appropriate expression in this internal language. This sort of explanation has frequently been attacked by philosophers, particularly Ryle (1949) and Wittgenstein (1953), as incomprehensible, but this is surely a case of a conceptual argument that fails when applied to computational models. Many computer systems have been built that have internal languages in this sense, and we are unable to find appeals to any features of human cognition in the usual arguments against internal languages that would make these arguments inapplicable to those systems. In particular, the internal language used in one of these computer systems always has a well-defined syntax, and usually a clear notion of inference defined in terms of manipulations of the formulas of the language. Whether these languages have truth-conditional semantics is more problematical, but for purposes of psychological explanation this may well be unnecessary. After all, if truth-conditional semantics cannot be given for the internal language the machine uses, although we know how to explain the behavior of the machine (since it was specifically designed to have that behavior), then such semantics cannot be required for the explanation. But if truth-conditional semantics is not required to perform "psychological explanation" for machines, why should it be required for humans?

Even if we accept the existence of computer systems that use such an internal language as a "model-theoretic" demonstration that the arguments against internal languages are misguided, just where they go wrong remains an interesting question. It would clearly be impossible to examine all such arguments in this brief chapter, (and we confess that we are not scholars of that literature,) but it may be instructive to look at at least one example. One familiar type of argument used by behaviorists against any number of concepts in cognitive psychology runs something like this:

> The only evidence admissible in psychology is behavioral evidence. There will always be many hypotheses, equally compatible with any possible behavioral evidence, about what X an organism has. Therefore, there is no empirical content to the claim that an organism has one X rather than another. Therefore the notion of X is unintelligible.

Quine has used this type of argument repeatedly in his discussions of the indeterminacy of translation (1960), ontological relativity (1971a), and knowledge of grammatical rules (1972). "Set of expressions in an internal language" is one of the concepts frequently substituted for "X" in this schema. The argument has some plausibility when applied to the human mind, where we have very little idea of how expressions in an internal language might be physically represented. It loses that plausibility when applied to computational models. If we recast the argument we can see why:

> The only evidence admissible for analyzing computer systems is behavioral evidence. There will always be many hypotheses, equally compatible with any possible behavioral evidence, about what set of expressions in an internal language form the basis of a computer system's "beliefs." Therefore, there is no empirical content to the claim that a computer system has one set of expressions rather than another. Therefore the notion of a set of expressions in an internal language in a computer system is unintelligible.

Where this argument breaks down depends on what is taken to be behavioral evidence. If we take behavioral evidence to be simply the input/output behavior of the system when it is running normally, then there is certainly more than behavioral evidence to draw upon. With a computer system we can do the equivalent of mapping out the entire "nervous system," and so understand its internal operations as well. On the other hand, if behavioral evidence includes internal behavior, it becomes much less plausible to say that there will be no way to tell which set of expressions the system possesses.

At this point, a computer scientist might be tempted to shout, "Of course! To find out what internal expressions the systems has, all you have to do is to print them out and look at them!"—but there is more to be said for the Quinean argument than this. What are directly observable, after all, are the physical states of the machine and their causal connections. There are many levels of interpretation between them and the print-out containing the set of expressions we wish to attribute to the machine. A Quinean might argue that there will be other interpretations that will lead to a different set of expressions, perhaps in a different internal language. With computer systems, however, the fact that they are designed to be interpreted in a certain way makes it extremely likely that any alternative interpretation would be far less natural, and so could be rejected on general grounds of simplicity and elegance. If this were not the case, it would be like discovering that the score of Beethoven's Ninth is actually the score of Bach's Mass in B Minor under a different, but no more complex, interpretation of the usual system of musical notation.

The Quinean argument fares somewhat better when applied to humans because there is no a priori reason to assume that human brains are designed to be interpreted in any particular way. Thus it is more plausible that there might be multiple descriptions of the operation of the brain in terms of internal languages, and that these descriptions, while incompatible with one another, are nevertheless equally compatible with all the evidence, including neurological evidence. But as the example of the computer system shows, and contrary to the Quinean argument, there is also no a priori reason to assume that this must be the case. It is, as the saying goes, an empirical question. It should be clear that one of the empirical commitments of any theory in cognitive psychology is that there be a preferred interpretation of the physical system in terms of the entities postulated by the theory. If this commitment is recognized, then failure to find a preferred interpretation makes the theory not incoherent or unintelligible, but simply false.

In view of all this, the best that can be said for the Quinean argument is that it points out the possibility that there will be more than one theory compatible with any evidence that can be obtained. But this is always the case in science. Surely no one would suggest that atomic theory is incoherent because there might be some as yet undiscovered alternative that is equally compatible with the evidence. Thus our consideration of computational models leads us to agree with Chomsky (1975, p. 182) that Quine's indeterminacy doctrine comes to no more than the observation that nontrivial empirical theories are underdetermined by evidence.

4.3 A Computational Model of Belief

The basic outlines of the computational model of belief presented below should be familiar to anyone acquainted with developments in artificial intelligence or cognitive simulation over the past few years. In calling this a model of belief, however, we must be careful to distinguish between psychology and semantics. Our model is intended to be a psychologically plausible account of what might be going on in an organism or system that could usefully be said to have beliefs. Even if we assume that the model does describe what is going on, the semantic question remains of how the English word "believe" relates to the model. We will put off addressing that question until Section 4.4.

As we said in the preceding section, belief will be explained in our model in terms of a system's being in a certain computational relation to expressions in an internal language. We will call the set of expressions to which a system is so related the *belief set* of the system.

The exact relationship between the expressions in this set and what we would intuitively call the beliefs of the system will be left unspecified until we discuss the semantics of belief sentences in Section 4.4. We will also be somewhat vague as to just what computational relation defines the belief set, but we can name some of the constraints it must satisfy. First of all, we will stipulate that, to be in the belief set of a system, an expression must be explicitly stored in the system's memory. It may turn out that we want to say the system has beliefs that would correspond to expressions that are not explicitly stored, but can be derived from stored expressions. In that case, the relationship between the system's beliefs and its belief set will be more complicated, but it will still be important to single out the expressions that are explicitly stored.

The fact that an expression is stored in the memory of the system cannot be sufficient, however, for that expression to be in the system's belief set. If the system is to be even a crude model of an intelligent organism, it will need to have propositional attitudes besides belief, which we would also presumably explain in terms of expressions in its internal language stored in its memory. We can account for this by treating the memory of the system as being logically partitioned into different spaces—one space for the expressions corresponding to beliefs, another for desires, another for fears, etc. These various spaces will be functionally differentiated by the processes that operate on them and connect them to the system's sensors and effectors. For example, perceiving that there is a red block on the table might directly give rise to a belief that there is a red block on the table, but probably not the desire or fear that there is a red block on the table. Similarly, wanting to pick up a red block might be one of the immediate causes of trying to pick up a red block, but imagining picking up a red block would presumably not.

This is a bit oversimplified, but not too much. Although it is true that perceiving a red block on the table could cause a fear that there is a red block on the table, this would need to be explained by, say, a belief that red blocks are explosive. In going from perception to belief, no such additional explanation is necessary. It seems completely compatible with our pretheoretical notions (which is what our model is supposed to reflect) to assume that we are simply built in such a way that we automatically accept our perceptions as beliefs unless they conflict with existing beliefs. (Anyone who does not think we are built this way should look out his window and try to disbelieve that what he sees is actually there.)

As to the internal language itself, we will again leave the details somewhat sketchy. For the purposes of this discussion, it will be sufficient to assume that the language is that of ordinary predicate logic augmented by intensional operators for propositional attitudes. The expressions in a belief set would be well-formed formulas in this language. The basic inference procedures should certainly be inclusive enough so that there is some way of applying them to generate any valid inference, but they could include procedures for generating plausible inferences as well. The important point is that, to interpret a set of formulas as a belief set, there had better be a well-defined notion of inference for them, since people clearly draw inferences from their beliefs. It is equally important, moreover, that there be a notion of an inference process in the model. The basic inference procedures merely define what inferences are possible, not what inferences will actually be drawn. There must be a global inference process that applies specific inference procedures to the formulas in the belief set and adds the resulting formulas to the belief set.

As simple as this model is, it seems to account fairly well for the obvious facts about belief. For example, it explains how "one-shot" learning can occur when one is told something. The explanation is that the hearer of a natural-language utterance decodes it into a formula in his internal language and adds the formula to his belief set. This idea, which seems to be almost universally accepted in generative linguistics and cognitive psychology, would hardly be worth mentioning if it were not for the fact that it differs so radically from the view presented in behaviorist psychology. According to standard behaviorist assumptions, we would expect that repeated trials and reinforcements would be necessary for learning to occur. This has some plausibility in the case of complex skills or large bodies of information, but a moment's reflection will show that very little learning fits this picture. Most "learning" consists of acquiring commonplace information such as where the laundry was put and what time dinner will be ready. Our model seems to explain this type of learning much better than does reinforcement of responses to a given stimulus.

A slightly less trivial, but still fairly obvious comment is that this model has no difficulty explaining how the system could accept one belief, yet reject another that is its logical equivalent. Suppose that beliefs are individuated more or less as are formulas in the internal language. Suppose further that the system has a particular formula P in its belief set that is logically equivalent to another formula Q, in the sense there is some way of applying the basic inference procedures of the system to infer Q from P and vice versa. The system may not put Q

in its belief set, however, because it never tries to derive Q, or because its heuristics for applying its inference procedures are not sufficient to find the derivation of Q, or because the derivation of Q is so long that it exhausts the system's resources of memory and time. We raise this point because the possibility that "A believes P" is true and "A believes Q" is false, even though P and Q are logically equivalent, is currently considered to be a major problem in the semantics of belief sentences, especially for theories based on possible-world semantics. In view of the voluminous literature this problem has generated (Montague 1974a; Partee 1973, 1979; Stalnaker 1976; Cresswell 1982), it is striking to note that, if reality is even vaguely like our a computational model, this is no problem at all for the psychology of belief. This suggests to us that the problem is artificial, a point we will return to in Section 4.5.

A more serious problem that can be handled rather nicely in this model is the question of what beliefs are expressed by sentences containing indexicals such as "I," "now," and "here." This is particularly troublesome for theories that take the language of thought to be identical to the external natural language. To take an example suggested by the work of Perry (1977, 1979), suppose that Jones has a belief he would express by saying "I am sitting down." We would take Jones's use of the word "I" to be a reference to Jones himself and take Jones's belief to be about himself. What is it that makes Jones's belief a belief about himself? It can't be simply that he has used the word "I" to express it, because he might not be using "I" as it is normally used in English; he must also believe or intend that in using "I" he refers to himself. But if this belief or intention consists in having certain English sentences stored in the appropriate space in his memory, it is hard to see how the explanation can avoid being circular. It is certainly not sufficient for Jones to believe "When I use 'I,' it refers to me," because this doesn't express the right belief or intention unless it has already been established that Jones uses "me" and "I" to refer to himself.

One way to try to get out of this problem is to say that Jones has some nonindexical description of himself and that his use of "I" is shorthand for this description. But, as Perry points out, having such a description is neither necessary nor sufficient to account for his use of "I." To see that it is not necessary, suppose Jones is the official biographer of Jimmy Carter, but he becomes insane and begins to believe that he actually is Jimmy Carter. Thus his beliefs include things he would express by "My name is 'Jimmy Carter,' " "I am President of the United States," "My daughter is Amy Carter," and so forth, in great detail. It does not seem to be logically impossible that *all* the nonindexical descriptions he attributes to himself are in fact true of Jimmy Carter and not true of him. On the description theory of indexicals, this should mean that Jones uses "I" to refer to Jimmy Carter and that his beliefs are all true. But it is intuitively clear that he still uses "I" to refer to himself, and that his beliefs are all delusional and false. On the other hand, suppose he is not insane, but uses "I" as a shorthand for some true description of himself such as "Jimmy Carter's biographer." Hence, when he says "I am sitting down" he expresses the belief "Jimmy Carter's biographer is sitting down." This does not explain his belief that he is sitting down, however, unless he also believes that *he* is Jimmy Carter's biographer.

In view of Kripke's (1972) critique of the description theory of proper names, it is not surprising that the description theory of indexicals doesn't work either. Nevertheless, it is interesting that Kripke's alternative, which does seem to work for proper names, still does not work for indexicals. Kripke's theory is essentially that when someone uses a proper name, it derives its reference from the occasion on which he acquired the use of the name, and that this creates a causal chain extending back to the original "dubbing" of the individual with that name. Thus, our use of "Kripke" refers to Kripke because we have acquired the name from occasions on which it was used to refer to Kripke. But this can't explain the use of the word "I," because no one *ever* acquires the use of "I" from an occasion on which it was used to refer to him.

In our computational model we can explain the use of "I" by assuming that the system has an individual constant in its internal language—call it SYS—that intrinsically refers to the system itself, and that the system uses "I" to express in English formulas of its internal language that involve this individual constant. This may seem to be no progress, since we are left with the task of explaining how SYS refers to the system. This is an easier task, however. A substantial part of the problem posed by "I" is that it is part of a natural language, and natural languages are acquired. The problem about beliefs being English sentences in the mind is that the person might have acquired a nonstandard understanding of them. Similarly, Kripke's causal-chain theory fails to explain the reference of "I" because "I" doesn't fit the assumptions the theory makes about how terms are acquired. As Fodor (1975) points out, however, if the internal language of thought is in fact not an external natural language, then we can assume that it is innate, and we are relieved of the problem of explaining how the expressions in it are acquired.

We can explain how SYS refers intrinsically to the system in terms of the functional role it plays. The system can be so constructed that, when it seems to see a red block, a formula roughly equivalent to "SYS seems to see a red block" is automatically added to the belief set, or at least becomes derivable in the belief space. Similarly, wanting to pick up a red block is intrinsically connected to "SYS wants to pick up a red block," and so forth. If the meaning of SYS is "hard-wired" in this way, then learning the appropriate use of "I" requires only learning something like "Use 'I' to refer to SYS." This type of explanation cannot be given in terms of the word "I" alone, because people are not hard-wired to use "I" in any way at all.

4.4 The Semantics of Belief Sentences

We hope the picture that we have presented so far is plausible as a model of the psychology of belief. If it is, then we have solved a number of interesting conceptual problems. That is, we have given at least a partial answer to the original question of how any physical system could have the properties that cognitive systems appear to have. Of course, solving conceptual problems is different from solving empirical problems; we have very little evidence that human cognitive systems actually work this way. On the other hand, we tend to agree with Fodor (1975, p. 27) that the only current theories in psychology that are even remotely plausible are computational theories, and that having remotely plausible theories is better than having no theories at all.

In light of the foregoing, there is a truly remarkable fact: although the psychology of belief is relatively clear conceptually, the semantics of belief sentences is widely held to suffer from serious conceptual problems. This might be less remarkable if the authors who find difficulties with the semantics of belief sentences rejected our conceptual picture, but that is not necessarily the case. For instance Cresswell (1982, p. 73) acknowledges that "it is probably true that what makes someone believe something is-indeed standing in an certain relation to an internal representation of a proposition," and it appears that Partee (1979) would also be favorably inclined towards this kind of approach.

It seems to us that, if we have a clear picture of what the psychology of belief is like, it ought to go a long way towards telling us under what condition attributions of belief are true. That is, it ought to give us a basis for stating the truth-conditional semantics of belief sentences. Our general view should be clear by now: if our computational model of belief is roughly the way people work, then "*A* believes that *S*" is true if and only if the individual denoted by "*A*" has the formula of his internal language that corresponds appropriately to "*S*" in his belief set, or can perhaps be derived in his belief set with limited effort. This latter qualification can be included or excluded, according to whether one wants to say that a person believes things he may never have thought about but that are trivial inferences from his explicit beliefs, such as the fact that 98742 is an even number, or that Anwar Sadat is a creature with a brain.

To complete this view we have to specify the relation between an attributed belief and the corresponding formula in the belief set. As a first approximation, we could say that "A believes that S" is true if and only if the individual denoted by "A" has in his belief set a formula he would express by uttering "S." For example, "John believes that Venus is the morning star" would be true if and only if the person denoted by "John" has a formula in his belief set that he would express as "Venus is the morning star." We believe this formulation is on the right track, but it has a number of difficulties that need to be repaired. For one thing, it is obviously not right for *de re* belief reports, such as "John believes Bill's mistress is Bill's wife." On its most likely reading, "Bill's mistress" is a description used by the speaker of the sentence, not John. We would not expect John to express his belief as "Bill's mistress is Bill's wife." We will return to the issue of *de re* belief reports later, but for now we will confine ourselves to *de dicto* readings.

Another apparent problem is the notion of a sentence in an external language expressing a formula of an internal language, but this can be dealt with by the same sort of functional explanation that we used initially to justify the notion of a belief set. A sentence expresses the internal formula that has the right causal connection with an utterance of the sentence. That causal connection may be complicated, but it is basically like the one between the contents of a computer's memory and a print-out of those contents. We will therefore assume that, given a causal account of how the production of utterances depends on the cognitive state of the speaker, there is a best interpretation of which formula in the internal language is expressed by a sentence in the external language.

A genuine problem in our current formulation is the fact that a person cannot be counted on to express his belief that Venus is the morning star as "Venus is the morning star," unless he is a competent speaker of English. A possible way around this would be to say that A believes that P if A has in his belief set a formula of his internal language that a competent speaker of English would express by uttering "S." This would be plausible, however, only if we assume that every person has the same internal language, and that expressions in the language can be identified across individuals. It might well be true that the internal language has the same syntax for all persons, since this would presumably be genetically determined, but that is not enough. We would have to further assume that a formula in the internal language means the same thing for every person.

This is clearly not the case, however, as many examples by Putnam (1975, 1977), Kaplan (1977), and Perry (1977, 1979) demonstrate. What these examples show is that two persons can be in exactly the same mental state (which, on our view, would require having the same belief sets), yet have different beliefs, because their beliefs are about different things. This should not be surprising, since there is nothing in our computational model to suggest that the reference or semantic interpretation of every expression in the language of thought is innate. Some expressions can be considered to have an innate interpretation because of the functional role they play in the model. Logical connectives and quantifiers in the internal language might have an a priori interpretation because of the way they are treated by innate inference procedures, and we have already discussed the idea that a cognitive system could have a constant symbol that intrinsically refers to the system. Predicates and relations for perceptual qualities, such as shapes and colors, would also seem to have a fixed interpretation based on the functional role they play in perception.

For most other expressions, including most individual constants and nonperceptual functions, predicates, and relations, there seems no reason to suppose that the interpretations are innately given. In fact, "concept learning" seems to be best accounted for by assuming that the internal language has an abitrarily large number of "unused" symbols on which information can be pegged. Acquiring a natural-kind concept might begin by noticing regularities in the perceptual properties of certain objects and deciding to "assign" one of the unused predicate symbols to that type of object. Then one could proceed to investigate the properties of these objects, adding more and more formulas involving this predicate to his belief set. Note that there is no reason to assume that the formulas added to the belief set constitute a biconditional definition of the concept; hence this picture is completely compatible with Wittgenstein's (1953) observation that we typically do not know necessary and sufficient conditions for application of the concepts we possess. Furthermore, since it is the acquisition process that gives the predicate symbol its interpretation, we can accommodate Putnam's (1975, 1977) point that a concept's extension can be partly determined by unobserved properties of the exemplars involved in its acquisition. This also demonstrates, contrary to Fodor (1975), that concept learning can be explained in terms of an internal language, without assuming that the language already contains an expression for the concept.

It appears that concept acquisition processes like the one suggested above could provide the symbols of the internal language with a semantic interpretation via the sort of causal chain that Kripke and Putnam discuss in connection with the semantics of proper names and natural-kind terms. Assuming the details can be worked out, we can use this semantic interpretation to try to define sameness of meaning across persons for expressions in the internal language. We can do this along lines suggested by Lewis's (1972) definition of meaning for natural languages: an expression P has the same meaning for A as Q has for B if P and Q have the same syntactic structure and each primitive symbol in P has the same intension for A as the corresponding symbol in Q has for B. We take an intension to be a function from possible worlds to extensions, and we assume that the intension of a primitive symbol is either innate, because of the functional role of the symbol, or is acquired in accordance with the causal-chain theory.

The problem with this definition is that two primitive symbols can have the same intension, but differ in what we would intuitively call meaning. Suppose John believes that Tully and Cicero are two different people. He might have in his belief set expressions corresponding to:

Name(Person3453) = "Tully"
Name(Person9876) = "Cicero"
Not(Person3453 = Person9876)

The best that the causal-chain theory can do for us is to provide the same intension for both Person3453 and Person9876, a function that picks out Cicero in all possible worlds. But clearly, these two symbols do not have the same meaning for John. In general, we probably would want to say that two symbols differ in meaning for an individual unless they have the same intension *and* are treated as such in the person's belief set (e.g., by having a formula asserting that they are necessarily equivalent).

To accommodate this observation, we will say that if the primitive symbol P has the same intension for A that the primitive symbol Q has for B, then P has the same meaning for A that Q has for B, providing either that these are the only symbols having that intension for A and B, or that the same expression in a common external language expresses P for A and Q for B. This latter condition may seem arbitrary, but it will allow us to say that if Bill and John both believe that Cicero denounced Catiline and Tully did not, then they both believe the same things. To use Quine's (1971a, p. 153) phrase, this amounts to "acquiescing in our [or in this case, Bill and John's] mother tongue."

These criteria obviously do not guarantee that, if two persons possess symbols with the same intension, there is some way to determine which ones have the same meaning. There may be other conditions that would allow us to do this that we have not thought of, but there will undoubtedly be residual cases. Suppose a language has two terms, P and Q, that, unknown to the speakers of the language, are rigid designators for the same natural kind, and so have the same intension. In a language that has only one term for this natural kind, it might well be impossible to express the belief that these speakers express when they say, "Some P's are not Q's." Imagine a culture in which the idea of the relativity of motion was so deeply embedded that they had no concept of X going around Y rather than Y going around X, but only X and Y being in relative circular motion. How would we go about explaining to them what it was that got Galileo into trouble?

We are finally in a position to state the truth conditions for *de dicto* belief reports that seem to follow from our computational model. First, we will say that an English expression "S" expresses the meaning of an internal expression P for an individual A just in case, for any competent English speaker B, there is an internal expression Q that has the same meaning for B as P has for A, and "S" expresses Q for B. Then a *de dicto* belief report of the form "A believes that S" is true if and only if the individual denoted by "A" has in his belief set a formula P such that "S" expresses the meaning of P for him.

To modify this theory to account for *de re* belief reports we will essentially reconstruct Kaplan's (1969) approach to apply to the internal language. According to our computational model, having a belief comes down to having the right formula in one's belief set, and a belief report tells us something about that formula. A *de dicto* belief report, such as "John believes Venus is the morning star," provides us with a sentence that expresses the meaning of the formula in the belief set. In a *de re* belief report, such as "John believes Bill's mistress is Bill's wife," part of the sentence, in this case "Bill's wife," need not express the meaning or intension of any part of the corresponding formula. Instead, it expresses the reference of part of the formula. Suppose that the relevant formulas in John's belief set are something like:

Name(Person55443) = "Bill"
Wife(Person55443) = Person12345

If these formulas are the basis for the assertion that John believes Bill's mistress is Bill's wife, then it must at least be the case that the occurrences of Person12345 in John's belief set refer to Bill's mistress. Otherwise, if John's belief is about anybody at all, then it is that person rather than Bill's mistress whom John believes to be Bill's wife. Something more than this is required, though. *De re* belief reports are generally held to support existential generalization. That is, from the fact that John believes Bill's mistress is Bill's wife we can infer that there is someone whom John believes to be Bill's wife. Phrasing it this way, however, we seem to be saying that John not only believes Bill is married, but he can pick out the person he thinks Bill is married to. If John has merely been told that Bill has been seen around lately with a beautiful woman and he has inferred that she must be his wife, then we could not really say that there is some specific person that he believes to be Bill's wife. There seems to be a certain amount of identifying information that John must have about Person12345 for his belief set to justify a *de re* belief report, although it is not always clear exactly what this information would be.

Now we can fully state our theory of the semantics of belief sentences. A sentence of the form "*A* believes *S*" is true if and only if the individual denoted by "*A*" has in his belief set a formula *P* that meets the following two conditions: first, the subexpressions of "*S*" that are interpreted *de dicto* must express the meaning for him of the corresponding subexpressions of *P*; second, the subexpressions of "*S*" that are interpreted *de re* must have the reference for him of the corresponding subexpressions of *P*, and he must be able to pick out the reference of those subexpressions of *P*.

4.5 Conclusion

The truth-conditional semantics for belief sentences presented above is a fairly complicated theory, but that really should not count against it. Most of its complexity was introduced to explain how a belief report in English could be true of someone who is not a competent speaker of that language. Most alternative theories of belief ignore this question entirely. All the formulations of possible-world semantics for belief that we know of, for instance, assume an unanalyzed accessibility relation between a person and the possible worlds compatible with his beliefs. That relation must surely be mediated somehow by his psychological state or his language, but no explanation of this is given. Furthermore, the most serious problem that plagues possible-world theories, the problem of distinguishing among logically equivalent beliefs of the same person, is no problem at all in our theory.

The really interesting question for us, though, is not whether one particular semantic theory is superior to another, but why so little effort has been made thus far to develop an account of the truth conditions of belief sentences in terms of psychological states and processes. Since belief is a psychological state, it seems that this would be the most natural approach to follow. Almost all the recent work on the semantics of belief sentences, however, appears to strive for independence from psychology. Most of this work tries to define belief in terms of a relation between persons and some sort of nonpsychological entities, with the relation either left unanalyzed or analyzed in nonpsychological terms (e.g., Hintikka 1962, 1971; Montague 1974b; Partee 1973, 1979; Stalnaker 1976; Cresswell 1982; Quine 1960, 1971b). We can only speculate as to why this is the case, but we can think of at least two probable motivations.

One motivation is what Cresswell calls "the autonomy of semantics" —the idea that the goal of semantics is to characterize the conditions under which a sentence of a language is true, and that this can be done independently of any considerations as to how someone could know what the sentence means or believe that what the sentence says is true. Thus we can say that "The cat is on the mat," is true if and only if the object referred to by "the cat" bears the relation named by "is on" to the object referred to by "the mat", without raising or answering any psychological questions. The point that the truth conditions of sentences do not in general involve psychological notions seems well taken, but it surely does not follow that they *never* do. No one seems to object to giving the truth conditions of sentences about physical states in terms of physical relations and physical objects, as in the example above. Why then, should there be any objection to giving the truth conditions for sentences about psychological states in terms of psychological relations and psychological objects?

To look at the matter a little more closely, the possible-world theories attempt to give the semantics of belief sentences in terms of *semantic* rather than psychological objects. That is, these theories claim that the objects of belief are built out of the constructs of the semantic theory itself. This would be a very interesting claim if it were true, but the failure up to now to make such a theory work suggests that it is probably not. If this assessment is correct, it seems natural to assume that the truth conditions of sentences about belief and other psychological states will involve the objects described by true psychological theories. If a true theory of the psychology of belief turns out to require the notion of an internal language, then it is probable that the truth conditions for belief sentences will involve expressions of that language.

The other motivation for seeking a nonpsychological semantics for belief sentences is the desire to unify the kind of truth-conditional semantics that we have been discussing with what is sometimes called "linguistic semantics," the task of characterizing what competent speakers know about the "meaning" of the sentences of their language. The most straightforward way to make this unification is to assume that the semantic knowledge that competent speakers of a language have is knowledge of the truth conditions of the language's sentences—a view that is, in fact, widely endorsed (Davidson 1967a, Moravcsik 1973, Partee 1979, Woods 1981, Cresswell 1982). It is quite implausible, however, that the kind of theory we have been sketching is what people know about belief or belief sentences. The root of the problem is our claim that the truth conditions for belief sentences can ultimately be stated only in terms of a true theory of the psychology of belief. But it is no more plausible that all speakers know such a theory than that all speakers know true theories of physics, chemistry, or any other science.

Our answer to this objection is that the idea that the semantic knowledge of speakers amounts to knowledge of truth conditions is simply mistaken. This is a general point that applies not only to sentences about psychological states, but to many other kinds of sentences as well. As we mentioned in Section 4.4, Putnam has convincingly argued that the extension of natural-kind terms generally depends not simply on what speakers of a language know or believe about the extension of the term, but also on what properties the objects that the term is intended to describe actually possess. But this means that speakers do not, in fact, know the truth-conditions of sentences that involve natural-kind terms. The properties that speakers believe characterize the extension of a natural-kind term may turn out to be incomplete or even wrong. When it was discovered that whales are mammals, what was discovered was just that. It was not discovered that whales did not exist, even if being a fish was previously central to what speakers of English believed about the truth conditions of "X is a whale." In general, the truth conditions for a natural-kind term depend not so much on the knowledge of competent speakers as on true scientific theories about the natural kind in question. Viewed from this perspective, the truth conditions of belief sentences depend on what turns out to be true in psychology because belief states form natural kinds in the domain of psychology.

According to our computational model, what a competent speaker of a language needs to know about the meaning of a sentence is not its truth conditions, but what formula in his internal language the sentence expresses in a given context. Of course, as we discussed in Section 4.4, this formula has truth conditions, and it seems plausible to say that the truth conditions of a sentence in a context are the same as those of the formula it expresses in that context. Now, knowing a formula in the internal language that has the same truth conditions as the sentence is something like knowing the truth conditions of that sentence, but not very much like it. In particular, it is nothing like knowing the statement of those truth conditions in any of the semantic theories we have discussed.

In the case of a belief sentence, the corresponding formula in the internal language might be thought of as an expression in a first-order language with a belief operator. If the hearer of "John believes that snow is white," takes "John" to refer to the same person as his internal symbol Person98765, and takes "snow is white" to express White(Snow), then the whole sentence might express for him the formula Believe(Person98765,White(Snow)). The functional roles and causal connections of the symbols in this formula determine its truth conditions, and those must be right for this formula to actually have the meaning for the hearer that John believes snow is white. Otherwise the hearer has not understood the sentence. To get those truth conditions right the hearer might have to have a lot of knowledge about belief, such as that people generally believe what they say, that they often draw inferences from their beliefs, and that they usually know what they believe. Knowing these properties of belief would help pin down the fact that belief is the psychological state denoted by Believe, yet these properties do not by any means constitute necessary and sufficient truth conditions for formulas involving Believe. But it is only required that these formulas have such truth conditions, not that the hearer *know* them.

The mistaken attempt to identify truth conditions with what speakers know about the meaning of sentences in their language has led to many pseudoproblems. For instance, Partee (1979) raises the question of whether for possible-world semantics to be correct, an infinite number of possible world models would have to exist in our heads. She concludes that they would not because "performance limitations" could let us get by with a finite number of finite models. This whole issue seems to be pretty much beside the point, however. Even for notions for which possible-world semantics appears to be adequate, such as the concept of necessity, nothing approximating possible worlds needs to be in our heads, although something like modal logic might.

Another example of the confusion that results from trying to unify these two notions of semantics is Woods's (1981) attempt to base a theory of meaning on "procedural semantics." Woods tries to identify the meaning of a sentence with some sort of ideal procedure for verifying its truth, saying that this procedure is what someone knows when he knows the meaning of the sentence. This has an advantage over possible-world semantics in that it can provide distinct meanings for logically equivalent sentences, since two different procedures could compute the same truth value in all possible worlds. The "procedures" that Woods is forced to invent, however, are not computable in the usual sense, even in principle. For example, to account for quantification over infinite sets he proposes infinite computations, while for propositional attitudes he suggests something like running our procedures in someone else's head. The sense in which these nonexecutable procedures are procedures at all is left obscure.

Partee starts from a particular notion of truth conditions, that of Montague semantics, and asks how such conditions could be represented in the head of a speaker. Woods starts from something that could be in the head of a speaker, i.e., procedures, and tries to make them yield truth conditions. In both cases, unlikely theories result from trying to say that it is truth conditions that are in the head, when all that is required is that what is in the head *have* truth conditions.

In this chapter we have examined a wide range of issues from the perspective of computational models of psychological processes and states. These issues include the legitimacy of psychological models based on internal languages, the problem of distinguishing logically equivalent beliefs, the psychology of having beliefs about oneself, belief reports about a nonspeaker of the language of the report, and the relation between truth-conditional and linguistic semantics. We do not claim to know whether the computational models we have proposed provide a correct account of all the phenomena we have discussed. What we do claim, however, is that many abstract arguments as to how things must be can be shown to be incorrect, and that many confusing conceptual problems can be clarified when approached from the standpoint of the concrete examples that computational models can provide.

5

Propositional Attitudes and Russellian Propositions

5.1 Introduction

An adequate theory of propositions needs to conform to two sets of intuitions that pull in quite different directions. One set of intuitions concerning entailments (or, more specifically, the *lack* thereof) among reports of propositional attitudes such as belief, knowledge, or desire points toward a very fine-grained notion of proposition. To be the objects of attitudes, propositions must seemingly be individuated almost as narrowly as sentences of a natural language. On the other hand, other intuitions seem to require that propositions not be specifically linguistic entities—rather that they be proper "semantic" objects, whatever that really amounts to.

Over the last few years, a number of approaches have been proposed in the attempt to reconcile these two types of intuitions. I believe that the simplest approach with any hope of success is the recent revival of the "Russellian" view of propositions, based on the semantic ideas expressed in *The Principles of Mathematics* (Russell 1903, Chapter V). Russell's idea at that time seems to have been that a proposition consists of a relation and the objects related. This contrasts with the "Fregean" view that a proposition must contain something like

This work was supported by a gift from the System Development Foundation. I would like to thank David Israel for his advice and criticism in the development of the ideas presented in this chapter, as well as for many helpful comments on the first draft. I would also like to thank Gordon Plotkin for pointing out that I was mistaken to claim, in the original version of the chapter, that it was impossible for all functions from individuals to propositions to be constituents of propositions, and for suggesting how the construction of propositions and propositional functions of finite type could be extended to higher ordinals so that all functions from individuals to propositions would, in fact, be constituents of propositions.

concepts of the objects related by the relation, rather than containing the objects themselves.

In this chapter, we explore the Russellian view of propositions and its adequacy as a basis for the semantics of propositional attitude reports. We review some of the familiar problems of attitude reports and suggest that a number of other approaches to their solution fall short of the mark. We then sketch how these problems can be handled by the Russellian approach, pointing out that it in fact offers a more complete treatment of the problems than is sometimes realized, and we present a formal treatment of a logic based on the Russellian conception of a proposition. Finally we discuss a number of remaining issues, including the need to distinguish propositional functions from properties and the problem of proper names in attitude reports.[1]

5.2 The Problem of Attitude Reports

Although it is familiar ground to quite literally everyone versed in the problems of providing a formal account of the semantics of natural language, it is useful to set the stage by reviewing why there exists a problem at all regarding the semantics of propositional attitude reports. At the most general level, the goal of formal semantics is to assign meanings systematically to expressions of a natural language. What meanings are taken to be varies from theory to theory, but the notion of truth or truth conditions always plays a central role. The reason for this is perhaps best put by Cresswell (1982), who points out that the semantic principle we can be most certain of is that if two sentences have different truth values (in the same context of use), they do not have the same meaning. What counts as a *systematic* assignment of meanings also varies, but it usually involves some version of the principle of compositionality: the meaning of an expression is a function of the meaning of its parts.[2]

Putting together these two basic semantic principles leads us to the

[1] Soames (1987) presents a theory that is similar in many ways to the one proposed here. For the most part the ideas expressed in this chapter were developed independently, but wherever ideas are taken directly from Soames, an explicit citation is given.

[2] This blurs a distinction between two kinds of meaning. Call $meaning_1$ the meaning an expression has on a particular occasion of use, and $meaning_2$ the function of features of occasions of use that determines what $meaning_1$ the expression has on a particular occasion. We will by and large ignore this distinction, since most theories, insofar as they make the distinction at all, assume that both $meaning_1$ and $meaning_2$ are compositional. It should be kept in mind, however, that other options do exist, such as holding that the $meaning_1$ of an expression is a function of the $meaning_2$ of its parts.

standard test for difference of meaning or (to use a less pretheoretically loaded term) semantic value: a difference must exist whenever substitution fails to preserve truth value. That is, suppose there is a sentence S containing an expression E, such that substituting E' for E in S results in a sentence S' that differs in truth value from S, in the same context of use. E and E' must then differ in semantic value. Why? Well, S and S' must differ in semantic value because they differ in truth value. Since the semantic values of S and S' must be a function of the semantic values of their parts, they must have some corresponding parts that differ in semantic value. But the only corresponding parts of S and S' that are not identical are E and E'; hence these must differ in semantic value.

Propositional-attitude reports provide a context for applying this test to proposals for the semantic value of whole sentences, since they contain whole sentences as proper parts. Attitude reports are typically of the form "*A V*'s that *S*," where *V* is an attitude verb such as "believe", "know", or "realize" and *S* is an embedded sentence; for example, "John knows that two plus two is four." The principle of compositionality requires that the semantic value of an attitude report be a function of the semantic value of the embedded sentence, and of no other feature of the embedded sentence.[3] Any assignment of semantic values to sentences, then, has to pass the substitution test with respect to propositional-attitude reports: if "*A V*'s that *S*" and "*A V*'s that *S*′" differ in truth value, then S and S' cannot be assigned the same semantic value.

In discussing attitude reports informally, one usually says that the embedded sentence expresses the proposition toward which the agent is reported to bear a certain attitude. The connection between this informal description of attitude reports and formal theories of their semantics is made by taking the semantic value of a sentence (on an occasion of use) to be the proposition it expresses (on that occasion). This identification may seem to beg an important question, but in practice the identification usually turns out to be stipulative; propositions are simply defined to be whatever the semantic theory assigns as the semantic value of sentences.

The major reason attitude reports are problematic is that using them in the substitution test seems to require a narrower individuation of propositions than any other phenomenon in language. If we were

[3]This assumes that the embedded sentence counts as a single part, or syntactic constituent, of the attitude report. It is considered to be so in just about every formal semantic theory, at least for so-called *de dicto* interpretations of attitude reports.

content to give a semantic account of a sufficiently restricted subset of a natural language, excluding propositional-attitude reports, among other things, it might suffice to take the semantic values of sentences to be simply their truth values. This is the approach adopted in the standard Tarskian semantics for first-order logic. The only sentential operators in the language of first-order logic are truth-functional, so there is no need to consider the semantic values of sentences as being anything more complex than their truth values.

As soon as we enrich the language, however, the Tarskian approach becomes inadequate. If we add modalities for necessity and possibility, we can create contexts in which substitution of embedded sentences with the same truth value does not preserve the truth value of the embedding sentence. For example, the sentence "It is necessary that two is an even number" is true and contains the embedded true sentence "Two is an even number." If we substitute for the latter the true sentence, "The earth is the third planet from the sun," however, the resulting larger sentence, "It is necessary that the earth is the third planet from the sun" is false. Hence we cannot assign the same semantic value to "Two is an even number" and "The earth is the third planet from the sun" without giving up at least one of our two basic semantic principles. The standard response to this type of example has been to take the semantic value of a sentence to be not its truth value, but the set of "possible worlds" in which it is true. The use of such notions in giving a formal semantics for natural language originated with Montague (1974c) and, until recently, has been far and away the dominant approach in the field. This approach handles the current example, since "Two is an even number" is true in all possible worlds while "The earth is the third planet from the sun" is not.

Possible-world semantics yields finer-grained semantic values of sentences than Tarskian semantics, but propositional-attitude reports require an even narrower individuation of semantic values of sentences in order to pass the substitution test. According to possible-world semantics, there is only one necessarily true proposition, the one that is true in all possible worlds, and, similarly, only one necessarily false proposition. But "A V's that S" and "A V's that S'" can differ in truth value even if S and S' are both necessarily true or both necessarily false. Consider a mathematician who does not believe some conjecture that is eventually proved to be a theorem. David Hilbert, for example, certainly did not believe at the time he posed his famous list of problems that first-order logic was undecidable, yet as a competent mathematician he obviously did believe many other necessary mathematical truths, such as that two is a square root of four. Thus "Hilbert believed that first-order logic is undecidable" is false, but "Hilbert believed that two is a square root of four" is true, although on the possible-worlds theory, "First-order logic is undecidable" and "Two is a square root of four" have the same semantic value, i.e., the proposition that is true in all possible worlds. The major problem posed by the semantics of propositional-attitude reports, then, is to find a plausible theory of propositions that individuates them narrowly enough to avoid making false predictions about entailments among attitude reports.

5.3 How Fine-Grained Must Propositions Be?

As Soames (1987) has pointed out, one can take the central idea of possible-world semantics—that the semantic value of a sentence is the set of possible worlds in which it is true—and separate it into two aspects. One aspect is that the semantic value of a sentence is the collection of circumstances that support its truth, while the other is that truth-supporting circumstances should be complete possible worlds. One way of viewing situation semantics, at least in its original version (Barwise and Perry 1983), is that it seeks to solve the problem of attitude reports by retaining the idea that the semantic value of a sentence is the collection of circumstances supporting its truth, but choosing a more finely individuated type of truth-supporting circumstances than possible worlds.

Situations are, in fact, exactly that. They are modeled as being more or less like partial possible worlds, determining the truth of certain atomic propositions, but having nothing to say about others. Hence, for any complete possible world, there would be many situations that are part of that world. Barwise and Perry use this device to try to get at the notion of a statement's subject matter. In their theory, a statement like "Joe is eating" does not entail "Joe is eating and Sarah is sleeping or Sarah isn't sleeping." There are situations that support the truth of the first but not the second, because they say nothing about whether or not Sarah is sleeping. This seems to solve the problem presented by the Hilbert examples. There are situations that support the truth of "Two is a square root of four," because they include the square-root relation holding between two and four. Many of these situations, however, say nothing about whether first-order logic has the property of decidability. The collection of situations that support the truth of "Two is a square root of four" is therefore distinct from the collection of situations that support the truth of "First-order logic is undecidable," so these two sentences express different propositions. By individuating truth-supporting circumstances more narrowly, Perry and Barwise manage to individuate the semantic values of sentences more narrowly.

It is not clear, however, that even replacing possible worlds with situations individuates propositions narrowly enough. Consider the following example: Suppose there is a building with a triangular cross section and three doors, A, B, and C, each equidistant from the other two. Suppose an agent wants to enter the building and so tries door A, but finds it locked. If he has no information about the other two doors, we would expect him to be indifferent as to which he should try next, since they are equidistant from him and from each other. However, if he knows that, whenever A is locked, B is not (and knows nothing else), we would expect him to go to door B, since that would ensure his getting into the building, whereas going to C would not.

Now, the interesting question is, what should he do if he tries door B first and finds it locked, knowing that, whenever A is locked, B is not? A moment's thought shows that he should go to door A, because if it is the case that, whenever A is locked, B is not, then it is also the case that, whenever B is locked, A is not. The point is that it does take a moment's thought to realize this; it is entirely possible that one could know that, whenever A is locked, B is not, and not realize that, whenever B is locked, A is not. Thus it is possible for "John knows (believes, realizes) that, whenever A is locked, B is not locked" to be true while "John knows (believes, realizes) that, whenever B is locked, A is not locked" is false.

This is clearly yet another counterexample to possible-worlds semantics (since "Whenever A is locked, B is not locked" and "Whenever B is locked, A is not locked" will be true in exactly the same possible worlds), but it may also be a counterexample to situation semantics. There is a problem, at least superficially, because both propositions appear to have the same subject matter; they involve exactly the same objects and properties. Unlike the mathematical example discussed previously, it seems that one cannot solve the problem by finding situations that support the truth of one of the sentences but, because they do not include the objects and relations the second sentence is about, do not support the truth of the other. Indeed, it seems difficult to imagine any independently motivated notion of truth-supporting circumstances that could distinguish between those that support "Whenever A is locked, B is not locked," and those that support "Whenever B is locked, A is not locked."

It would be hard to demonstrate categorically that the example under discussion cannot be treated successfully in situation semantics, because, in its unfinished state, one cannot say with certainty how "whenever" would be treated (but see the discussion in Section 5.5). Nevertheless, our example should raise doubts as to whether any approach based on distinguishing propositions by their truth-supporting circumstances can individuate propositions finely enough to avoid making false predictions concerning entailments among attitude reports.[4]

5.4 Could Propositions Be Syntactic?

The examples we have looked at suggest that propositions will have to be individuated almost as finely as sentences of a natural language to avoid undesirable entailments among attitude reports. This has led to the development of a number of theories of the attitudes as relations to syntactic objects. The simplest theory of this sort is that the objects of attitude reports are just sentences of a natural language. Quine (1971b) has suggested this approach, and Stalnaker (1984) has taken a position toward belief in mathematical propositions that is a slight variant of this: propositions are treated as sets of possible worlds, but mathematical propositions are interpreted as being about the truth of mathematical sentences. The fact that it is hard to imagine an agent who lacks language having sophisticated mathematical beliefs makes it at least somewhat plausible that mathematical beliefs are beliefs about sentences, but any "metalinguistic" approach of this general sort also raises problems. Most notably, it makes it difficult to explain what we are doing when we report in one language the beliefs of an agent who speaks a different language. The problem can be seen in the Hilbert examples. Hilbert's native language was German, so it is highly unlikely that when he thought about mathematics, he thought about the truth of English sentences. Hence the truth of "Hilbert believed that two is a square root of four" does not seem to depend on Hilbert's having any attitude at all toward the embedded sentence "Two is a square root of four," which is, after all, a sentence of English and not German. If we try to get around this problem by saying that "Hilbert believed that two is a square root of four" means Hilbert stood in the belief relation to a sentence of German whose English translation is "Two is a square root of four," we seem to beg the question, since the notion of translation appears to depend on the notion of sameness of meaning, and it is the difficulty of individuating meanings adequately that prompted the syntactic approach in the first place.

Even if this problem could be circumvented—perhaps by giving

[4]Soames (1987) comes to much the same conclusion, but his major argument derives from views that we do not accept concerning the truth conditions of attitude reports containing proper names. (See Section 5.8.)

a nonsemantic account of the translation relation—the metalinguistic approach does not seem to be applicable to all situations in which truth-supporting circumstances do not individuate propositions finely enough. The difference between believing that, whenever A is locked, B is not, and believing that, whenever B is locked, A is not, seems to have nothing to do with natural language. One could easily imagine an intelligent but non-language-using agent displaying behavior that one would associate with the first belief but not the second. That is, he might go immediately from door A to door B when he is unable to get through A, but might hesitate, or even go to door C, if he approached B first and found it locked. Although it might turn out that only language-using agents display this sort of behavior and that, whenever any other organism is taught something analogous to "Whenever A is locked, B is not locked," it also learns the corresponding analogue of "Whenever B is locked, A is not locked," this would be a significant empirical discovery (and a very surprising one at that), rather than a prediction that should follow logically from an a priori theory of the objects of the attitudes.

Problems such as these, as well as the popularity of the "representational theory of mind," have led to a related view that the semantics of propositional-attitude reports should be defined in terms of internal representations of the meanings of the sentences used to describe the content of the attitude. That is, what constitutes the belief that, whenever A is locked, B is not, is not an attitude toward an English sentence, but rather possessing in one's "belief store" an internal representation encoding the belief. The reason that one can believe that, whenever A is locked, B is not, without believing that, whenever B is locked, A is not, is that these have different representations in the internal language. This general picture is perhaps most closely associated with Fodor (1975), but the most explicit attempt to construct a semantics for belief reports based on it is probably one developed in part by this author (see Chapter 4).

This sort of theory resolves at least some of the problems of the metalinguistic approach to attitude reports. Once one accepts the notion of a system of internal representation, it is not implausible to assume that all humans have the *same* system. The sharing of a belief among speakers of different languages is accounted for by assuming that, while they may express the belief differently in their respective external languages, they have the same encoding of the belief in their common internal language. Similarly, a non-language-using agent, if he has complex beliefs, may be assumed to have an internal language in which those beliefs are encoded. Two beliefs that are true in exactly the same circumstances may be different because they have different encodings in the agent's system of internal representation, even though the agent lacks an external language to express them.

All this may well be true, but it is still not clear whether the postulation of a system of internal representation solves the problems of the metalinguistic approach or merely postpones them. Even if all people have similar systems of internal representation, it does not seem very plausible that all agents capable of holding propositional attitudes do. A dog digs a hole in the lawn because he believes there is a bone buried there. The dog's human owner, observing him, comes to hold the same belief. Must the dog and the human have the same system of internal representation for them both to be able to believe there is a bone buried at a particular spot on the lawn? In this case, it seems that something more abstract than syntactic structure is needed to attribute the same belief to both dog and human.

There is another problem in treating the semantics of propositional-attitude reports that is particularly troublesome for syntactic theories of the attitudes: the problem of quantifying into propositional-attitude contexts. If quantifiers range over actual objects like chairs, tables, and people, then an attitude report like "There is someone whom Ralph believes to be a spy" is difficult to interpret on a syntactic theory of the attitudes. The quantified sentence is true just in case some actual person has the property of being believed by Ralph to be a spy. But this seems to require that the person be in some sense a constituent of what Ralph believes and, if the objects of beliefs are syntactic entities, it is hard to see how this could be. In a formal treatment of a syntactic approach to the attitudes (e.g., Konolige 1985), a substitutional theory of quantification can be adopted, along with the assumption that the system of representations that encode beliefs includes a unique standard designator for each element of the domain. This assumption makes the logic function smoothly, but is wildly implausible as a theory of the attitudes of real agents. Attempts to describe more plausible accounts of quantifying-in that are compatible with a syntactic theory of the attitudes invariably become very complicated and seem to require a treatment of quantification into attitude reports that is quite different from the treatment of other types of quantification. (Kaplan's (1969) early attempt to explain quantifying-in illustrates nicely the complexities that arise.[5]) The need to provide a simple and uniform treatment of

[5] Although Kaplan was working within a Fregean theory of propositional attitudes rather than a syntactic theory, the central problem in both cases is that the entities over which quantification range cannot be constituents of the objects of the attitudes.

quantifying into attitude contexts, along with related problems raised by pronouns in attitude reports, perhaps as much as anything else, points toward a Russellian treatment of the objects of the attitudes.

5.5 The Russellian Theory

Compared with the other approaches we have discussed, the basic Russellian picture of propositions is striking in its simplicity. A proposition is simply a structured entity composed of an n-ary property or relation and n objects. A proposition is true if and only if the objects stand in the relation, and propositions are identical if and only if they consist of the same objects standing in the same relation in the same way.[6] It might be tempting to regard a Russellian proposition as being a *sequence* of a relation and its arguments. It seems preferable, though, to regard propositions as a distinct sort of object in their own right, and, although one might wish to model them as sequences, there is no reason to take them literally to be sequences.

Since the Fregean view that propositions must contain concepts of objects rather than the objects themselves has been dominant for something like eighty or ninety years, the idea that propositions have things like tables, chairs, and people as constituents may seem strange. There is an intuition that physical objects exist in one realm and abstract objects in another, and that it doesn't make sense to claim that objects from one side of this divide have objects from the other side as constituents. However, a moment's thought reveals that set theory has exactly the same problem as soon as we go beyond the universe of purely mathematical objects and consider sets of ordinary objects, like the set of all the artifacts in the British Museum. Sets are abstract objects, but this set contains such things as statues, clocks, coins, and jewelry. So this is an example of an abstract object that contains physical objects, yet we do not find it in any way strange. Alternatively, we could sidestep the issue by remaining agnostic as to whether propositions literally contain objects (or, for that matter, properties and relations) and simply hold that propositions are individuated by their subject matter. Hence, if we change the relation or any of the objects the proposition is about, we obtain a different proposition.

Because they are structured objects, Russellian propositions can be made to mirror the structure of sentences closely enough to distinguish propositions that may be true in the same circumstances. Yet they are

[6]By "the same way" we mean, roughly speaking, in the same order. That is, the proposition that John is taller than Bill must be distinguished from the proposition that Bill is taller than John.

in no sense syntactic or linguistic objects, since they are defined only in terms of objects, properties, and relations. They are consequently free from the problems of language specificity that plague the sorts of theories discussed in Section 5.4.

It seems questionable whether Russell himself ever presented a satisfactory theory of the nature of logically complex propositions. In his writings where the "Russellian" notion of a proposition is most clearly articulated, only the simplest sorts of propositions are treated in a way that would at all appeal to a modern semanticist. In Chapter V of *Principles of Mathematics*, Russell says of the proposition that Socrates is human that it contains Socrates and the property of being human. The only types of more complex propositions he considers in that discussion are those that involve quantification, but the theory of quantification he presents there is certainly not something one would wish to adopt.

The most straightforward way of extending Russell's ideas about elementary propositions to handle more complex ones is to treat propositions and propositional functions as objects that can themselves enter into propositions as arguments of properties and relations. In particular, logical connectives such as conjunction, disjunction, and negation can be treated as relations among, or properties of, propositions, while quantifiers can be treated as relations among, or properties of, propositional functions.

To explore this approach, it will be useful to introduce some formal notation for propositions. Given the simplicity of the theory, the notation can also be extremely simple. A Russellian proposition will be denoted simply by a parenthesized sequence consisting of an expression denoting an n-ary property or relation followed by expressions denoting the n objects being related. Thus the expression $(\mathsf{R}\ \mathsf{A}_1 \ldots \mathsf{A}_n)$ denotes the Russellian proposition that the relation denoted by R holds among the objects denoted by $\mathsf{A}_1, \ldots, \mathsf{A}_n$, and, in accord with the individuation conditions for propositions, the expression $(\mathsf{R}\ \mathsf{A}_1 \ldots \mathsf{A}_n)$ denotes the same proposition as the expression $(\mathsf{R}'\ \mathsf{A}'_1 \ldots \mathsf{A}'_{n'})$ if and only if R and R' denote the same relation, n is equal to n', and, for every i from 1 through n, A_i denotes the same object as A'_i.

Our examples about doors being locked can be handled by regarding "whenever" as expressing a relation between propositions, and "not" as expressing a property of propositions. Thus "A is locked" expresses the proposition that A has the property of being locked, (Locked A); "B is not locked" expresses the proposition that the proposition that B is locked has the property of not being the case, (Not (Locked B)); and "Whenever A is locked, B is not locked" expresses the proposition that the proposition that A is locked and the proposition that B is not locked stand in the relation of the second's being the case whenever the first is the case,

(Whenever (Locked A) (Not (Locked B))).

This last proposition is clearly different from the proposition that, whenever B is locked, A is not,

(Whenever (Locked B) (Not (Locked A))),

because, in the two cases, the "whenever" relation holds between different pairs of propositions.[7]

As we mentioned above, quantification can be handled by treating quantifiers as properties of, or relations among, propositional functions. We can, for example, interpret "some" and "every" as expressing generalized quantifiers that relate two propositional functions. If we use lambda expressions to denote propositional functions, "Every man is mortal" could then be interpreted as expressing the proposition

(Every (λx (Man x)) (λy (Mortal y))).

This is the proposition that the "every" relation holds between the function that maps an individual into the proposition that the individual is a man and the function that maps an individual into the proposition that the individual is mortal; that is, it is the proposition that every individual that satisfies the first propositional function also satisfies the second.[8]

Just as in the examples with "whenever", many quantified propo-

[7]There is, as is often the case, a problem here with respect to time and tense. We might not want to consider the embedded sentences in "Whenever A is locked, B is not locked" as expressing complete propositions because they are not specific as to time. The received wisdom is that propositions must be true or false absolutely, not relative to a time. If we wanted to hold to that position, we would have to treat the semantic values of the embedded sentences not as propositions, but as functions from times to propositions. It is not so clear, however, which are the objects of propositional attitudes. The conventional wisdom, again, is that the objects of the attitudes are absolute rather than time-relative, but consider the following case: In 1952, John formed the belief that Nixon was a crook and has held to that opinion ever since. Naively, this seems like one belief that he has maintained for over thirty years. But the object of that belief would then not be temporally absolute; it could have been false in 1952, but true in 1972. The alternative that is actually compatible with the conventional wisdom would be that at every instant, he has a new belief that Nixon is a crook at that instant—rather counterintuitive. Whichever way we come down on this issue, it is not difficult to adjust the semantic theory accordingly. Hence we will follow a long-established tradition and ignore time and tense in the remainder of this chapter.

[8]We will say that an object satisfies a propositional function if and only if the function maps the object into a true proposition.

sitions that are logically equivalent will nonetheless be distinct. The proposition that not every man is mortal,

(Not (Every (λx (Man x)) (λy (Mortal y)))),

will be distinct from the proposition that some man is not mortal,

(Some (λx (Man x)) (λy (Not (Mortal y)))),

because the first says that the proposition that every man is mortal has the property of not being the case, whereas the second says there is something that satisfies the function that maps an individual into the proposition that the individual is a man and also satisfies the function that maps an individual into the proposition that the individual is not mortal.

The Russellian approach thus provides, in a very natural way, more finely individuated semantic values for sentences than do any of the approaches we have considered that are based on collections of truth-supporting circumstances. Moreover, it allows for a straightforward treatment of quantification into attitude contexts, which was a major difficulty with syntactic approaches to propositional-attitude reports. To be consistent with the examples given above, we would want to say that "Some man is believed by Ralph to be a spy" expresses the proposition

(Some (λx (Man x)) (λy (Believe Ralph (Spy y)))).

That is, some individual who satisfies the propositional function (λx (Man x)) also satisfies the propositional function (λy (Believe Ralph (Spy y))). This is the point at which things came unstuck in the syntactic approach to attitude reports. In order to satisfy the propositional function (λx (Man x)), an object must have the property of being a man. That is, it must be an actual flesh-and-blood person. To satisfy (λy (Believe Ralph (Spy y))), however, the object must be such that the belief relation holds between Ralph and the proposition that the object is a spy. On a syntactic (or a Fregean) approach, it is not clear what this proposition would be, since actual individuals cannot be constituents of beliefs. At best, concepts or expressions that *denote* individuals can be constituents of beliefs, so some complicated story has to be told about how to get from an actual individual to the right sort of concept or expression. Moreover, one is faced with the unpalatable choice of either telling the same complicated story with respect to *all* quantification or treating quantification into attitude reports very differently from other kinds of quantification.

On a Russellian approach there are no such complications. The theory of quantification is both simple and uniform. An individual satisfies the function (λx (Man x)) just in case the property of being a man holds of the individual. An individual satisfies the function (λy (Believe Ralph (Spy y))) just in case the belief relation holds between Ralph and the proposition that the property of being a spy holds of the individual. The treatment is the same in both cases because there is no difficulty in having actual indivduals as constituents of propositions.

It is worth emphasizing that Russellian propositions seem to be able to provide *all* the objects needed for an account of the semantics of attitude reports. In the paper that, perhaps more than any other, led to the revival of interest in Russellian propositions, Kaplan (1977)[9] treats them as an adjunct to, rather than a substitute for, Fregean propositions. Kaplan's view at that time seemed to be that some expressions do have both a sense and a denotation, just as Frege maintained, but that other expressions (particularly indexicals and demonstratives) have only a denotation, so that all they can contribute to the proposition expressed by the sentence that contains them is the objects they denote. Kaplan seemed to want to retain Fregean propositions (which he calls "general" propositions) where only expressions of the former type were involved, but to allow Russellian propositions (which he calls "singular" propositions) as well to handle expressions of the latter type.

It is easy to see why it might seem that having Russellian propositions does not dispense with the need for Fregean propositions. Consider the sentence "The father of Bill is happy." What does the noun phrase "the father of Bill" contribute to the proposition the sentence expresses? According to the commonsense view that this phrase denotes Bill's father, it seemingly must contribute something other than just its denotation because, even if Bill has the same father as Mary, "The father of Bill is happy" and "The father of Mary is happy" appear to express different propositions. One could easily be in a position to believe one but not the other. This is of course just the problem that led Frege to make the distinction between denotation and sense, and something like Fregean senses seem to be needed for its solution.

How would belief reports be analyzed according to Kaplan's picture? If we take the belief report "John believes that the father of Bill is happy," we note an ambiguity, traditionally described in terms of the *de dicto/de re* distinction. On the *de re* interpretation, we would be saying of Bill's father that John believes him to be happy, which we would analyze in terms of John's believing a Russellian singular propo-

[9]After circulating in typescript form for many years, Kaplan's paper was eventually published in a volume of papers from a Kaplan festschrift conference (Almog, Perry, and Wettstein 1989, pp. 481-614).

sition that includes the property of being happy and Bill's father. On the *de dicto* interpretation, we would be saying that John believes a Fregean general proposition that includes the property of being happy and the Fregean sense of the noun phrase "the father of Bill." Russellian propositions are needed to account for *de re* interpretations of belief reports, but Fregean propositions seem to be needed to account for *de dicto* interpretations.

This really does seem to have been Kaplan's view (1977, p. 13–15). Kaplan even notes (1977, footnote 9), and suggests a solution to, a problem that arises when both Fregean and Russellian propositions are allowed within the same theory. The problem is to distinguish a general proposition that contains the Fregean sense of an expression from a singular proposition about that Fregean sense itself. For instance, how do we distinguish the Fregean proposition that the father of Bill is happy from the Russellian proposition that literally says that the sense of "the father of Bill" is happy? On Kaplan's view they both have the same structure and the same constituents.

Kaplan's proposed solution to this problem need not concern us here because, by taking a different approach to the interpretation of complex noun phrases, we can eliminate the need for Fregean senses and avoid the problem altogether. The approach in question is basically just the one contained in Russell's (1949) famous theory of descriptions: treat all complex noun phrases as quantified noun phrases. If we treat "the" as a generalized quantifier instead of a singular-term-forming operator, the problem raised by Frege goes away.[10] "The father of Bill is happy" would be taken to express the proposition that the propositional function expressed by "father of Bill" and that expressed by "is happy" stand in the relation expressed by "the":

$$(\mathsf{The}\ (\lambda \mathsf{x}\ (\mathsf{Father}\ \mathsf{x}\ \mathsf{Bill}))\ (\lambda \mathsf{y}\ (\mathsf{Happy}\ \mathsf{y}))).$$

This will of course be distinct from the proposition that the propositional function expressed by "father of Mary" and that expressed by "is happy" stand in the "the" relation, as well as from the overtly singular proposition that simply attributes the property of being happy to a particular man who happens to be the father of both Bill and Mary. It is not necessary that the relation expressed by "the" be the particular one that would be consistent with Russell's theory of descriptions (or that it be viewed as structurally complex, which is a source of many objections to Russell's theory). So long as it is some generalized quan-

[10]Except for proper names, which are dealt with in Section 5.8.

tifier, the Russellian theory of propositions will not fall prey to Frege's counterexamples.

Kaplan (1977, p. 15) asserts that, when Russell propounded his theory of descriptions, he gave up the theory of propositions that we are elaborating. Whether or not he did discard that theory is a question we can leave to the Russell scholars; the point to be made here is that he need not have done so. Russell's theory of the meaning of noun phrases certainly changed dramatically between 1903 and 1905. But nothing in the later theory is inherently incompatible with the basic picture of the structure of propositions presented in the earlier work.

Before going on to consider some of the more subtle issues connected with the Russellian theory of propositions, it is worth taking one more look at the possibilities for successfully treating attitude reports within the framework of situation semantics. In Section 5.3 we doubted that an approach based on treating situations as fine-grained truth-supporting circumstances would provide a solution to the problem of attitude reports, but we cautioned against assuming that no solution within the general framework of situation semantics could be found. Indeed, recent work on situation semantics (Barwise and Perry 1985, Barwise and Etchemendy 1987) suggests that the problem of distinguishing "Whenever A is locked, B is not locked" and "Whenever B is locked, A is not locked" might be attacked by treating "whenever" as a relation between states of affairs,[11] so that a situation would support the truth of "Whenever A is locked, B is not locked" just in case it includes the "whenever" relation holding between the "A is locked" state of affairs and the "B is not locked" state of affairs.

This would indeed distinguish the two sentences in question, since, from a formal standpoint, there is nothing to require a situation that includes the "whenever" relation holding between the "A is locked" and "B is not locked" states of affairs to also include the "whenever" relation holding between the "B is locked" and "A is not locked" states of affairs, or vice versa. To adopt this treatment, however, is really to abandon the idea that situations are truth-supporting circumstances. Suppose we have an actual situation that really does completely establish the truth of "Whenever A is locked, B is not locked," e.g., one that, for every point in time, settles the questions of whether A is locked and whether B is locked and, for every point for which it includes A's being locked, also includes B's not being locked. Such a situation would not *formally* support the truth of "Whenever A is locked, B is not locked" unless it also included the "whenever" relation holding between the "A

[11] Which were called "situation-types" in the original version of the theory.

is locked" state of affairs and the "B is not locked" state of affairs. Furthermore, if a situation does include this, it does not matter what else it contains. It seems, then, that we can dispense with all the elements of the theory that allow us to treat situations as truth-supporting circumstances, and simply identify the semantic content of "Whenever A is locked, B is not locked" with a single semantic object that consists of the "whenever" relation holding between the "A is locked" state of affairs and the "B is not locked" state of affairs. This, however, is no longer an alternative to the Russellian notion of a proposition; it simply *is* that notion.[12]

5.6 Russellian Logic

In this section we formalize the syntax and semantics of the notation introduced in the preceding section. We can think of this as giving us a kind of "Russellian logic" that could play the same role in providing a Russellian semantics for fragments of natural language that Montague's intensional logic has played with respect to possible-world semantics.

One of the main problems in formalizing Russellian logic stems from our decision to use lambda expressions to denote propositional functions. We want the propositional functions in our theory to be just what they seem to be, namely, functions from objects to Russellian propositions—functions that could be modeled as sets of ordered pairs of objects and propositions.[13] If we were to allow arbitrary lambda expressions in our logical language, we would face the well-known problem that the unrestricted lambda calculus does *not* have simple models in terms of functions viewed as sets of ordered pairs in standard set theory. All the issues of concern in this chapter, however, arise even if we restrict our attention to first-order quantification, which, within our framework, amounts to dealing only with propositional functions that map individuals into propositions. This restricted case can be treated relatively simply, so we will accept the restriction and confine our attention to functions from individuals to propositions.

[12] Barwise and Etchemendy (1987) advocate what they call "Austinian" propositions, which they distinguish from Russellian propositions in that they describe a particular situation rather than the entire world. Both types of propositions they present, however, are the kinds of structured semantic objects we are considering; consequently, from standpoint of the issues discussed in this chapter, both are Russellian.

[13] It is perhaps an obvious point, but it is worth emphasizing that they need be nothing more than functions-in-extension. We do not require anything intensional in a Russellian logic, because Russellian propositions already offer a sufficiently rich domain of extensions to make all the distinctions we need.

These considerations lead us to define a first-order Russellian language to be a language with the following categories of expressions:

- Predicate constants, or simply predicates, which are atomic symbols denoting properties and relations.
- Individual terms, which are either individual variables or individual constants.
- Functional terms, or lambda expressions, which denote propositional functions and are of the form $(\lambda \mathsf{v}\ \mathsf{F})$, where v is an individual variable and F is a formula.
- Propositional terms, or formulas, which denote propositions and are of the form $(\mathsf{R}\ \mathsf{A}_1 \ldots \mathsf{A}_n)$, where R is an n-ary predicate and $\mathsf{A}_1, \ldots, \mathsf{A}_n$ are terms.

Well-formed expressions will be said to be *closed* if every individual variable they contain is bound within some lambda expression according to the usual notion of bound variable.

There are a number of observations about this language that may be warranted. First, formulas are just one type of term rather than a fundamentally different category. They are simply terms that denote propositions. We do not even consider it ill-formed to attribute to a proposition what we would normally think of as a property of an individual. We could, for instance, say that the proposition that two is a square root of four is taller than John; it would merely be a falsehood. Treating formulas as terms that denote propositions enables us to introduce propositional attitudes just as relations between agents and propositions. Similarly, there are no special provisions made for logical operators or quantifiers, since these are treated simply as particular properties of, and relations among, propositions and propositional functions. Finally, it should be noted that although we have introduced lambda expressions to denote propositional functions, we have not introduced any notation to indicate functional application. We could easily do so, but it is not necessary for addressing the problems of interest in this chapter.

An interpretation of a first-order Russellian language is a quadruple $\langle \mathbf{I},\mathbf{R},\mathbf{E},\mathbf{C} \rangle$, where $\mathbf{I}$ is the domain of individuals, $\mathbf{R}$ is the domain of properties and relations, $\mathbf{E}$ is the function that maps each member of $\mathbf{R}$ into its extension, and $\mathbf{C}$ is the function that gives the denotation of the constants of the language, mapping each individual constant of the language into a member of $\mathbf{I}$ and each predicate constant of the language into a member of $\mathbf{R}$.

To provide denotations for formulas and lambda expressions, we also need a domain of propositions and a domain of propositional functions, but these can be defined in terms of **I** and **R**. A system of propositions and propositional functions of finite type suffices for this purpose, so we define by mutual induction **P**, the domain of propositions of finite type, and **F**, the domain of propositional functions of finite type, as follows:

- Let $\mathbf{P}_i$ be the smallest set such that both:
  - For all $j < i$, $\mathbf{P}_j$ is a subset of $\mathbf{P}_i$.
  - For every n, every n-ary relation R in **R**, and every sequence $A_1, \ldots, A_n$ of members of $\mathbf{I} \cup \mathbf{P}_i \cup \mathbf{F}_i$, the proposition that R holds among $A_1, \ldots, A_n$ is an element of $\mathbf{P}_i$.
- Let $\mathbf{F}_0$ be the empty set.
- Let $\mathbf{F}_{i+1}$ be the set of all functions from **I** into $\mathbf{P}_i$.
- Let $\mathbf{P} = \bigcup_{i=0}^{\infty} \mathbf{P}_i$.
- Let $\mathbf{F} = \bigcup_{i=0}^{\infty} \mathbf{F}_i$.

We can think of $\mathbf{P}_i$ as the set of propositions of type i and $\mathbf{F}_i$ as the set of functions of type i from individuals to propositions. The domains **P** and **F**, then, are respectively the set of all propositions of finite type, and the set of all functions of finite type from individuals to propositions, that can be constructed from the set of individuals **I** and the set of properties and relations **R**. Having defined **P** and **F**, we can now characterize **E** a bit more precisely: if R is an n-ary property or relation in **R**, $\mathbf{E}(R)$ is the set containing each n-tuple $\langle A_1, \ldots, A_n \rangle$ of members of $\mathbf{I} \cup \mathbf{P} \cup \mathbf{F}$, such that R holds among $A_1, \ldots, A_n$.

The domains **P** and **F** are not the largest domains of propositions and propositional functions we could define, but as we will see below, they are adequate to provide denotations for all the formulas and lambda expressions in our logic. In some sense, **F** is restricted more than **P**. Relative to **I**, **R**, and **F**, every proposition that could reasonably be in **P** according to the Russellian picture *is* in **P**. The only way to generate more propositions would be to have more individuals, more properties and relations, or more propositional functions. **F**, however, contains far fewer than all the functions from **I** into **P**. It contains only those that can be assigned a finite type. If we wish, however, we can enlarge **P** and **F** so that **F** does contain all the functions from **I** into **P**, simply by generalizing the construction of $\mathbf{P}_i$ and $\mathbf{F}_i$ to higher ordinals.[14]

The construction goes almost exactly as before. We simply substitute variables ranging over ordinals for variables over natural numbers to get definitions of $\mathbf{P}_\alpha$ and $\mathbf{F}_\alpha$ for any ordinal α, except that, if α is

[14]This was pointed out by Gordon Plotkin.

a limit ordinal, $\mathbf{F}_\alpha$ is defined to be $\bigcup_{\beta<\alpha} \mathbf{F}_\beta$. The definitions of **P** and **F** in the original construction, then, would simply give us $\mathbf{P}_\omega$ and $\mathbf{F}_\omega$ in the extended construction. It is easy to show that, in the extended construction, the size of $\mathbf{P}_\alpha$ and $\mathbf{F}_\alpha$ are bounded by a cardinal that depends on the size of **I** and **R**. Since there are more ordinals than can be put in one-to-one correspondence with any cardinal, and $\mathbf{F}_{\alpha+1}$ must always be at least as big as $\mathbf{F}_\alpha$, it follows that there must be some ordinal α such that $\mathbf{F}_{\alpha+1} = \mathbf{F}_\alpha$, which in turn means that $\mathbf{P}_{\alpha+1} = \mathbf{P}_\alpha$. Inspection of the definitions of $\mathbf{P}_\alpha$ and $\mathbf{F}_\alpha$ shows that this is a level where $\mathbf{F}_\alpha$ contains all the functions from **I** into $\mathbf{P}_\alpha$, and that $\mathbf{P}_\alpha$ and $\mathbf{F}_\alpha$ will remain the same at all higher ordinals. Either this extended construction or the original construction limited to finite types can be assumed in the rest of the chapter without any significant differences.

With respect to an interpretation $\langle \mathbf{I},\mathbf{R},\mathbf{E},\mathbf{C} \rangle$, we can define the denotation of every closed well-formed expression and the truth of every closed formula. We define a two-place partial function **D** such that, if g is a partial function from individual variables to their values and E is a well-formed expression, $\mathbf{D}(g, \mathsf{E})$ is the denotation of E under the variable assignment g. $\mathbf{D}(g, \mathsf{E})$ will be undefined if E contains a free occurrence of a variable that is not assigned a value by g. We can define **D** as follows:

- If E is an individual constant or predicate constant, then $\mathbf{D}(g, \mathsf{E}) = \mathbf{C}(\mathsf{E})$.
- If E is an individual variable, then $\mathbf{D}(g, \mathsf{E}) = g(\mathsf{E})$.
- $\mathbf{D}(g,(\lambda \mathsf{v}\ \mathsf{F}))$ is the function that maps each member A of **I** into $\mathbf{D}(g[\mathsf{v}/A], \mathsf{F})$, where $g[\mathsf{v}/A]$ is identical to g except that it gives the variable v the value A.
- $\mathbf{D}(g,(\mathsf{R}\ \mathsf{A}_1 \ldots \mathsf{A}_n))$ is the proposition that the relation $\mathbf{D}(g, \mathsf{R})$ holds among the objects $\mathbf{D}(g, \mathsf{A}_1), \ldots, \mathbf{D}(g, \mathsf{A}_n)$. Note that the identity conditions on Russellian propositions imply that $\mathbf{D}(g,(\mathsf{R}\ \mathsf{A}_1 \ldots \mathsf{A}_n)) = \mathbf{D}(g,(\mathsf{R}'\ \mathsf{A}'_1 \ldots \mathsf{A}'_{n'}))$ if and only if $\mathbf{D}(g, \mathsf{R}) = \mathbf{D}(g, \mathsf{R}')$, $n = n'$, and, for all i between 1 and n, $\mathbf{D}(g, \mathsf{A}_i) = \mathbf{D}(g, \mathsf{A}'_i)$.

The denotation of a well-formed expression E will be $\mathbf{D}(g_0, \mathsf{E})$, where g_0 is the partial function from variables to values that is undefined everywhere. This definition has the consequence that denotation is defined only for closed well-formed expressions.

Note that every closed formula has a denotation in **P** and every closed lambda expression has a denotation in **F**. Let the rank of a formula or lambda expression E be the maximum depth of embedding of lambda expressions in E. A simple inductive argument shows that, if

E is of rank n and $\mathbf{D}(g, \mathsf{E})$ is defined, then, if E is a formula, $\mathbf{D}(g, \mathsf{E})$ is a proposition of type n contained in $\mathbf{P}$, or, if E is a lambda expression, $\mathbf{D}(g, \mathsf{E})$ is a propositional function of type n contained in $\mathbf{F}$. It is also worth noting that, if all constants have distinct denotations, two closed well-formed expressions will always have distinct denotations unless they can be made identical by renaming bound variables. Thus we have achieved the goal of individuating propositions almost as finely as sentences without invoking any overtly syntactic notions.

The truth of a closed formula is easily defined in terms of $\mathbf{D}$ and $\mathbf{E}$. A closed formula $(\mathsf{R}\ \mathsf{A}_1 \ldots \mathsf{A}_n)$ is true if and only if $\langle \mathbf{D}(g_0, \mathsf{A}_1), \ldots, \mathbf{D}(g_0, \mathsf{A}_n)\rangle$ is in $\mathbf{E}(\mathbf{D}(g_0, \mathsf{R}))$; that is, if and only if the n-tuple of the denotations of $\mathsf{A}_1, \ldots, \mathsf{A}_n$ is in the extension of the denotation of R. It may seem that this does not say very much about the truth of formulas when compared with truth definitions for standard logics; we might have expected to see recursive clauses that give the truth conditions of complex formulas in terms of the embedded formulas they contain. In the general case, however, we do not want the semantic framework itself to impose any constraints on what the truth conditions of a complex proposition might be. Those truth conditions are simply specified by whatever $\mathbf{E}$ assigns as the extensions of the properties and relations used to build up a complex proposition. This is of particular concern with respect to propositional-attitude relations, such as belief. There might be empirical facts about the phenomenon of belief that would introduce constraints on the collection of propositions an agent believes, but logic itself should impose none.

On the other hand, we may wish to treat quantifiers and logical connectives as a special case among properties of, and relations among, propositions and propositional functions. When we introduce "logical" predicate constants such as Every and Not, we may wish to place extra conditions on $\mathbf{E}$ so that formulas involving these constants will be guaranteed to have their usual truth conditions. For instance, we can specify that Not actually represents negation by imposing on $\mathbf{E}$ the constraint that

$$\langle \mathbf{D}(g, (\mathsf{R}\ \mathsf{A}_1 \ldots \mathsf{A}_n))\rangle \in \mathbf{E}(\mathbf{C}(\mathsf{Not}))$$

if and only if

$$\langle \mathbf{D}(g, \mathsf{A}_1), \ldots, \mathbf{D}(g, \mathsf{A}_n)\rangle \notin \mathbf{E}(\mathbf{D}(g, \mathsf{R})).$$

That is, the property of not being the case holds of a proposition if and only if the principal relation of the proposition does not hold among its arguments. For Every, the relevant constraint is that

$$\langle \mathbf{D}(g, (\lambda \mathsf{v}\ (\mathsf{R}\ \mathsf{A}_1 \ldots \mathsf{A}_n))), \mathbf{D}(g, (\lambda \mathsf{v}'\ (\mathsf{R}'\ \mathsf{A}'_1 \ldots \mathsf{A}'_{n'})))\rangle \in \mathbf{E}(\mathbf{C}(\mathsf{Every}))$$

if and only if, for every A in $\mathbf{I}$ such that

$$\langle \mathbf{D}(g[\mathsf{v}/A], \mathsf{A}_1), \ldots, \mathbf{D}(g[\mathsf{v}/A], \mathsf{A}_n)\rangle \in \mathbf{E}(\mathbf{D}(g[\mathsf{v}/A], \mathsf{R})),$$

it is also the case that

$$\langle \mathbf{D}(g[\mathsf{v}'/A], \mathsf{A}'_1), \ldots, \mathbf{D}(g[\mathsf{v}'/A], \mathsf{A}'_{n'})\rangle \in \mathbf{E}(\mathbf{D}(g[\mathsf{v}'/A], \mathsf{R}')).$$

In other words, the "every" relation holds between two propositional functions just in case every individual that the first propositional function maps into a true proposition is also mapped into a true proposition by the second propositional function.

Clearly, similar constraints could be imposed on **E** for all the other usual logical particles. In fact, if we replaced the generalized quantifiers Every and Some with two one-place predicates of propositional functions, Forall and Exists, and we treated (∀v F) and (∃v F) as "syntactic sugar" for (Forall (λv F)) and (Exists (λv F)), respectively, we could furnish a Russellian semantics for the standard notation for first-order predicate logic that would be completely equivalent to standard model theory in terms of the truth conditions assigned to closed formulas. With a Russellian semantics, however, propositional-attitude relations can be accommodated simply by adding new predicates to the language and their denotations and extensions to the interpretation, without any undesirable entailments being created.

5.7 Why Propositional Functions?

One striking fact about the theory we have presented is the care with which we have distinguished properties and relations from propositional functions. It seems an obvious question to ask whether we could not dispense with one of these notions in favor of the other. After all, with both properties and propositional functions all one has to do to obtain a proposition is to add an argument in the right way. In fact, we could dispense with properties and relations in favor of propositional functions, or vice versa, but either change to the theory would have rather unpleasant consequences. In one direction the reduction would leave open important questions that are settled by the current theory; in the other direction it would introduce new questions for which there do not seem to be obvious answers.

If we tried to eliminate properties and relations in favor of propositional functions, we would no longer be able to say anything in general about the individuation conditions for propositions. Reinterpreting the notation (P A) so that it denotes the result of applying the propositional function denoted by P to the object denoted by A, we could no longer state that (P A) never denotes the same proposition as (P′ A′)

unless P and P′ have the same denotation and A and A′ have the same denotation. It might turn out that P denotes the same propositional function as (λx (R x A′)) and P′ denotes the same propositional function as (λy (R A y)), in which case (P A) and (P′ A′) would both denote the same proposition as (R A A′), no matter what R, A, and A′ denote. This would amount to taking take propositions as primitive, rather than properties and relations. Thomason (1980) has previously proposed taking propositions as primitive, which certainly does avoid the undesired entailments among attitude reports. It gives up two very intuitive consequences of the Russellian theory, however; (1) that propositions have a definite structure, and (2) that they are individuated by their subject matter.

The other possible reduction would be to eliminate propositional functions by interpreting lambda expressions as denoting complex properties. On this approach, the proposition that all men are mortal might still be denoted by

(Every (λx (Man x)) (λy (Mortal y))),

but (λx (Man x)) and (λy (Mortal y)) would be taken to denote properties, rather than propositional functions, and Every would denote a relation between properties. The individuation conditions on propositions would be retained, so that ((λx (R x A′)) A) and ((λy (R A y)) A′)[15] would denote distinct propositions, since they do not attribute the same property to the same object. They would, however, have the same truth conditions.

The main problem with this approach is that there seems to be no intuitive basis for settling many questions of identity for properties denoted by lambda expressions. For example, what is the relation between the property denoted by Man and that denoted by (λx (Man x))? If they are different, wherein lies the difference? If they are the same, by what principle can we draw that conclusion? Or, to take a slightly more complex example, we might consider the lambda expression (λxy (R y x)) to denote the converse of a relation R. The converse of the converse of R should be R itself, but this would require (λxy ((λxy (R y x)) y x)) and R to denote the same relation. What general principle would give us that identity?

The approach we have chosen avoids these problems. Man and (λx (Man x)) have different denotations because the first denotes a

[15]These would be ill-formed expressions in our current logic, but, if lambda expressions denoted properties and relations instead of propositional functions, it would be a completely arbitrary restriction not to allow them to appear in the predicate position of formulas.

property, the second a propositional function; they are of fundamentally distinct semantic categories. All questions about the identity of the denotations of lambda expressions are answered by the extensional individuation conditions for propositional functions. These will ensure that two lambda expressions denote the same function if and only if, roughly speaking, they have the same structure with all predicate and individual constants in corresponding positions having the same denotation, subject to renaming of variables.[16]

Of course, there are reasons other than the problems central to this chapter for developing a full-blown property theory with complex expressions denoting properties (e.g., nominalization), and the issues raised here might have to be addressed in that context. The problem of individuating propositions finely enough for a treatment of propositional attitudes need not raise those issues, though, so it seems best to adopt an approach that leaves them to one side.

5.8 Proper Names

We have left until last what is perhaps the most difficult problem of all concerning the semantics of attitude reports, the problem of proper names. Up to this point, we have assumed that the semantic value of a proper name is simply what we ordinarily regard as its referent. For example, in the sentence "John is happy," the semantic value of "John" would be some particular person named "John", the semantic value of "is happy" would be the property of being happy, and the semantic value of the whole sentence would be the Russellian proposition that attributes the property of being happy to the person named "John". The trouble with this treatment of proper names is that it seems possible to derive different propositions by substituting one proper name for another with the same referent. The classic example is that it seems that "Cicero" and "Tully" must have different semantic values, even though they are names for the same person, because "John believes that Cicero was a Roman orator" could be true while "John believes that Tully was a Roman orator" is false.

Recently the view that the semantic value of a proper name is always just its referent has come to be fairly widely held. Barwise and Perry (1983), Kaplan (1977), and Soames (1987) have all been among the advocates of this position, which has come to be called the "direct-reference theory". According to these theorists, "John believes that

[16] This assumes that functional application is not expressible. If we extended our language to permit functional application, then two lambda expressions would denote the same function if and only if their Church-Rosser normal forms met this condition.

Cicero was a Roman orator" actually does entail "John believes that Tully was a Roman orator," but there is a pragmatic reluctance to substitute "Tully" for "Cicero" because we tend to make belief reports do double duty—not only reporting what propositions are believed, but also indicating what sentences are accepted. Thus, "John believes that Cicero was a Roman orator" not only reports that John believes the proposition that Cicero/Tully was a Roman orator, but it also leads us to expect that John would accept as true the sentence "Cicero was a Roman orator." The reason we are reluctant to substitute "Tully" for "Cicero" is that it would be misleading, not that it would be false.

If arguments of this sort were correct, it would be most convenient for our Russellian theory of propositions, because the problem of proper names would be explained away and no modification of the theory would be required. Unfortunately, there are other examples that seem to reveal not merely a reluctance to substitute, but also instances of substitution that definitely lead from truth to falsehood. Suppose that someone comes to Stanford to meet the famous John Perry. Being of sound mind, and having followed Perry's work through the philosophy journals, he certainly knows (the necessarily true proposition) that John Perry is John Perry. When he arrives at Stanford, he goes directly to the Center for the Study of Language and Information, where a seminar is in progress, and asks the group "Which of you is John Perry?" One of the members of the Center turns to Perry and says "He does not know that you are John Perry." The direct-reference theory would predict that this last statement is simply false. Under the circumstances described, "You are John Perry" and "John Perry is John Perry" would express exactly the same proposition; hence the proposition expressed by "He knows that John Perry is John Perry" would be the same as "He knows that you are John Perry." Yet, in this case, our intuitions seem clear that the first is true and the second is false.

Positive arguments for the view that proper names are directly referential—that is, that their semantic value is simply their referent—are actually rather hard to find. In most of the work on direct reference, one mainly finds rather careful arguments that certain uses of pronouns are directly referential, followed by little more than a claim that proper names work the same way.[17] What this analogy does not

[17]There may be a mistaken reliance on Kripke's (1972) arguments that proper names are rigid designators. Kripke's arguments are quite convincing, at least with respect to modal contexts, and directly referential expressions would, a fortiori, be rigid designators, but this is merely consistent with the hypothesis that proper names are directly referential; it does not actually show that they are.

take into account is that proper names are subject to the same *de dicto/de re* distinction as definite descriptions, whereas pronouns normally are not. Recall that the belief report "John believes that the father of Bill is happy" is subject to an ambiguity as to whether the semantic content of the phrase "the father of Bill" is part of the belief attributed to John or is used to pick out a certain individual whom John has a belief about. Paraphrases that suggest one or the other of these interpretations would be "John believes that whoever is the father of Bill is happy," for the first, and "John believes of the father of Bill that he is happy," for the second. This same ambiguity exists if we substitute a proper name for the definite description. "John believes that Cicero was a Roman orator" has two interpretations that we could paraphrase by "John believes that whoever was Cicero was a Roman orator" and "John believes of Cicero that he was a Roman orator." The first of these is like a *de dicto* interpretation of a definite description, while the second is like a *de re* interpretation.

Similar sentences containing pronouns, on the other hand, appear to have only one interpretation, which behaves like the *de re* interpretation of a definite description or a proper name. Thus, speaking (across the centuries) to Cicero, we could say "John believes that you were a Roman orator." This seems to be much like saying "John believes of you that you were a Roman orator;" i.e., the *de re* interpretation. This is just what one would expect from a direct-reference treatment of pronouns, since only the referent of the pronoun would be available to be part of the semantic content of the utterance. Since sentences with proper names can also have *de re* interpretations, there is a basis for the claimed parallelism between proper names and pronouns. However, note that, on a *de re* interpretation, substitution of coreferential proper names is allowable. That is, the proposition expressed by "John believes of Cicero that he was a Roman orator" does entail that expressed by "John believes of Tully that he was a Roman orator." What the direct-reference theory does not account for is the apparent *de dicto* interpretations of proper names, for which substitution is invalid. The proposition expressed by "John believes that whoever was Cicero was a Roman orator" certainly does not entail that expressed by "John believes whoever was Tully was a Roman orator."

How, then, are we to treat the *de dicto* interpretation of proper names in attitude reports? It seems to me that a limited appeal to the metalinguistic approach provides the most satisfactory account. That is, on the *de dicto* interpretation, "John believes Cicero was a Roman orator" should be treated, roughly speaking, as if it were "John believes that the entity called 'Cicero' was a Roman orator." This will express a different proposition from "John believes that the entity called 'Tully' was a Roman orator," because the names "Cicero" and "Tully" actually form part of the propositions expressed. We argued in Section 5.4 that the metalinguistic approach could not provide a general solution to the problem of individuating objects of the attitudes. Limited to the case of proper names, though, the arguments against the metalinguistic approach fail to apply. The principal argument was that the metalinguistic approach does not explain how we can report in a given language the attitudes of agents who are not speakers of that language. But to claim that "John believes that whoever was Cicero was a Roman orator" is true, while "John believes that whoever was Tully was a Roman orator" is false, it seems that John must be a speaker of a language that includes both the name "Tully" and the name "Cicero"; hence, the translation problem does not arise.

There is one slight twist, however, that makes things more complicated. While it seems that an agent must be a speaker of a language in which a certain name occurs in order to have a *de dicto* attitude involving the name, the name need not occur in the agent's language in exactly the form it does in the language of the attitude report. Thus we could report that John believes that Tully was a Roman orator, even if John's only language were Latin and he used "Tullius" instead of "Tully". Note, however, that "Tullius" and "Tully" are in a historiolinguistic sense the same name, rather than being two distinct names that in some mysterious way have the same meaning over and above their reference.

Kripke's (1979) famous example of Pierre who believes the city called "London" is ugly but that the city called "Londres" is beautiful relies on the transformation of a name in going from one language to another. This might initially seem to be a counterexample to a metalinguistic treatment of proper names, but in fact it is not. The crucial point is that in describing Pierre's beliefs in Kripke's example, we must resort to talking about the city called "London" and the city called "Londres" because we lack any clear intuitions as to whether we should say that "Pierre believes that London is beautiful" is true relative to a *de dicto* interpretation of "London". This is exactly what we should expect if a metalinguistic treatment of proper names were correct, with the differing forms of the same name in different languages being treated as equivalent. If we think of Pierre as a speaker of French, we should say that he does believe that London is beautiful; if we think of him as a speaker of English, we should say that he does not believe that London is beautiful. Since he is presented as bilingual, our conventions for interpreting belief reports do not resolve the question.

A metalinguistic interpretation of *de dicto* occurrences of proper names in attitude reports seems, on the whole, to give the best explanation of our intuitions, but how should we fit it into the overall framework we are developing? Should we simply regard the metalinguistic interpretation as providing the standard semantic value for proper names? I think that we probably should not; we should rather regard it as a pragmatic reinterpretation that is applied to make sense out of problematical cases, with the direct-reference account actually providing the normal interpretation. We are not retreating, though, to the position taken by the "pure" direct-reference theorists. Their view was that the semantic value of proper names remains the same in all contexts, but that there are pragmatic restrictions on the use of certain attitude reports that might cause hearers to draw mistaken inferences. This can explain why speakers refrain from making certain utterances that would, according to their theory, express true propositions, but it cannot explain why speakers make other utterances that, according to their theory, express false propositions. The view advocated here, on the other hand, is that pragmatic considerations actually cause us to assign a nonstandard semantic value to a proper name when a *de dicto* interpretation of the proper name is required, which gives us a proposition that is in fact true in those cases that cause trouble for the pure direct-reference view.

There are several reasons for preferring this way of looking at things over a theory that says that proper names are always treated metalinguistically, but two considerations seem particularly compelling. One is the fact that the same problem arises with terms that denote kinds, accompanied by strong intuitions that the metalinguistic interpretation is exceptional. That is, we are not even aware of the metalinguistic interpretation unless the direct-reference interpretation is in some way odd. For example, in the sentence "John believes that oculists charge too much for their services," we do not sense any ambiguity in the interpretation of "oculists". On the other hand, "John believes that oculists are devil worshipers" suggests an ambiguity between an interpretation on which John has confused the words "oculist" and "occultist" and the more direct interpretation that John has a strange belief about the religious practices of eye doctors. The second reason for regarding the metalinguistic interpretation of proper names as a nonstandard alternative interpretation is that treating it as the standard interpretation yields wrong results in contexts other than attitude reports, particularly modal contexts. As Kripke (1972) has forcefully argued, proper names act as rigid designators in modal contexts, but a phrase such as "the entity called 'Cicero'" is not a rigid designator. It refers to whatever is called "Cicero", and that can obviously vary from one possible set of circumstances to another.

5.9 Conclusion

In this chapter, we have looked at what seem to be the major issues involved in developing a Russellian theory of propositions that is adequate to deal with the semantics of attitude reports. We have tried to sketch solutions to the most important problems, but in most areas there is clearly more work to be done. In particular, the treatment of proper names seems both the least elegant and least well-developed part of the theory. The theory is no worse than any other in this regard, however, since no one seems to have offered a better treatment of proper names within any framework.

Another issue that deserves more attention is the question of whether we may have individuated propositions *too* finely. As we remarked in Section 5.6, if all constants have distinct denotations, then every pair of closed formulas of our Russellian logic that cannot be made identical by renaming variables denote distinct propositions. If we decide that this is too strict a standard of propositional identity, there are various ways we could choose to relax it. For example, it is a commonly held view that, in conjunctions and disjunctions, the order of conjuncts or disjuncts should not affect propositional identity. That is, (And P Q) should denote the same proposition as (And Q P). We have suggested in Section 5.5 that conjunction and disjunction be treated as relations on propositions, but it would not conflict with the basic structure of the theory to treat them as properties of sets of propositions, or even as associative commutative operators in an algebra of propositions.

The theory is clearly in need of generalization beyond the restriction to first-order, well-typed propositional functions. The theory as it stands is an odd mixture, with completely untyped predication but, because of the role played by propositional functions, highly restricted quantification. It would clearly be no problem to extend the theory to higher-order, well-typed propositional functions, and hence to higher-order quantification. This would not, however, give us the ability to quantify over all propositions, since there is no type that encompasses them all. Use of models for the type-free lambda calculus (Stoy 1977) could perhaps provide a way to remove this restriction. Allowing predication over all propositions would of course make the theory vulnerable to the semantic paradoxes, but, given the current interest in solving the paradoxes rather than avoiding them, along with the fact that they do arise in natural language, any theory within which the issues cannot at least be raised should probably be considered inadequate.

Finally, although we have presented a formal language whose interpretation is given in terms of the Russellian theory of propositions developed here and have suggested informally what meaning, in terms of the theory, should be assigned to various sentences of English, a formal semantics for a significant fragment of English within the theory nonetheless remains to be presented. It seems clear that a semantics for such a fragment could be produced, but undoubtedly many interesting new issues and problems would be uncovered in the process.

Part III

Autoepistemic Logic

6

Semantical Considerations on Nonmonotonic Logic

6.1 Introduction

Commonsense reasoning is "nonmonotonic" in the sense that we often draw, on the basis of partial information, conclusions that we later retract when we are given more complete information. Some of the most interesting products of recent attempts to formalize nonmonotonic reasoning are the nonmonotonic logics of McDermott and Doyle (1980; McDermott 1982). These logics, however, all have peculiarities that suggest they do not quite succeed in capturing the intuitions that prompted their development. In this chapter we reconstruct nonmonotonic logic as a model of an ideally rational agent's reasoning about his own beliefs. For the resulting system, called *autoepistemic logic*, we define an intuitively based semantics for which we can show autoepistemic logic to be both sound and complete. We then compare autoepistemic logic with the approach of McDermott and Doyle, showing how it avoids the peculiarities of their nonmonotonic logic.

It has been generally acknowledged in recent years that one important feature of ordinary commonsense reasoning that standard logics fail to capture is its *nonmonotonicity*. An example frequently given to illustrate the point is the following. If we know that Tweety is a bird, we will normally assume, in the absence of evidence to the contrary, that Tweety can fly. If, however, we later learn that Tweety is a pen-

The research reported herein was supported by the Air Force Office of Scientific Research under Contract No. F49620-82-K-0031. The views and conclusions expressed in this document are those of the author and should not be interpreted as necessarily representing the official policies or endorsements, either expressed or implied, of the Air Force Office of Scientific Research or the U. S. Government. This chapter previously appeared in *Artificial Intelligence*, Vol. 25, No. 1, 1985, and is reprinted here with the permission of the publisher, Elsevier, Amsterdam.

guin, we will withdraw our prior assumption. If we try to model this in a formal system, we seem to have a situation in which a theorem P is derivable from a set of axioms A, but is not derivable from some set A' that is a superset of A. The set of theorems, therefore, does not increase monotonically with the set of axioms; hence this sort of reasoning is said to be "nonmonotonic." As Minsky (1974) has pointed out, standard logics are always monotonic, because their inference rules make every axiom *permissive*. That is, the inference rules are always of the form "P is a theorem if $Q_1, \ldots, Q_n$ are theorems," so that new axioms can only make more theorems derivable; they can never invalidate a previous theorem.

Recently there have been a number of attempts to formalize this type of nonmonotonic reasoning. The general idea is to allow axioms to be restrictive as well as permissive, by employing inference rules of the form "P is a theorem if $Q_1, \ldots, Q_n$ are *not* theorems." The inference that birds can fly is handled by having, in effect, a rule that says that, for any X, "X can fly" is a theorem if "X is a bird" is a theorem and "X cannot fly" is not a theorem. If all we are told about Tweety is that he is a bird, we will not be able to derive "Tweety cannot fly"; consequently, "Tweety can fly" will be inferable. If we are told that Tweety is a penguin and we already know that no penquin can fly, we will be able to derive the fact that Tweety cannot fly, and so the inference that Tweety can fly will be blocked.

One of the most interesting embodiments of this approach to nonmonotonic reasoning is McDermott and Doyle's "nonmonotonic logic" (1980; McDermott 1982). McDermott and Doyle modify a standard first-order logic by introducing a sentential operator "M," whose informal interpretation is "is consistent." Nonmonotonic inferences about birds being able to fly would be sanctioned in their system by the axiom (McDermott 1982, p. 33)

$$\forall x(\mathsf{Bird}(x) \wedge M(\mathsf{Can\text{-}Fly}(x)) \supset \mathsf{Can\text{-}Fly}(x)).$$

This formula can be read informally as "for all X, if X is a bird and it is consistent to assert that X can fly, then X can fly." McDermott and Doyle can then have a single general nonmonotonic inference rule, whose intuitive content is "MP is derivable if $\neg P$ is not derivable."

McDermott and Doyle's approach to nonmonotonic reasoning seems more interesting and ambitious than some other approaches in two respects. First, since the principles that lead to nonmonotonic inferences are explicitly represented in the logic, those very principles can be reasoned about. That is, if P is such a principle, we could start out believing $Q \supset P$ or even $MP \supset P$, and come to hold P by drawing inferences, either monotonic or nonmonotonic. So, if we use McDermott and Doyle's representation of the belief that birds can fly, we could also represent various inferences that would lead us to *adopt* that belief. Second, since they use only general inference rules, they are able to provide a formal semantic interpretation with soundness and completeness proofs for each of the logics they define. In formalisms that use content-specific nonmonotonic inference rules dealing with contingent aspects of the world (i.e., it might have been the case that birds could not fly), it is difficult to see how this could be done. The effect is that nonmonotonic inferences in McDermott and Doyle's logics are justified by the meaning of the premises of the inferences.

There are a number of problems with McDermott and Doyle's nonmonotonic logics, however. The first logic they define (1980) gives such a weak notion of consistency that, as they point out, MP is not inconsistent with $\neg P$. That is, it is possible for a theory to assert simultaneously that P is consistent with the theory and that P is false. McDermott subsequently (1982) tried basing nonmonotonic logics on the standard modal logics T, S4, and S5. He discovered, however, that the most plausible candidate for formalizing the notion of consistency that he wanted, nonmonotonic S5, collapses to ordinary S5 and is therefore monotonic. In the rest of this chapter we develop an alternative formalization of nonmonotonic logic that shows why these problems arise in McDermott and Doyle's logics and how they can be avoided.

6.2 Nonmonotonic Logic and Autoepistemic Reasoning

The first step in analyzing nonmonotonic logic is to determine what sort of nonmonotonic reasoning it is meant to model. After all, nonmonotonicity is a rather abstract *syntactic* property of an inference system, and there is no a priori reason to believe that all forms of nonmonotonic reasoning should have the same logical basis. In fact, McDermott and Doyle seem to confuse two quite distinct forms of nonmonotonic reasoning, which we will call *default reasoning* and *autoepistemic reasoning*. They talk as though their systems were intended to model the former, but they actually seem much better suited to modeling the latter.

By default reasoning we mean the drawing of plausible inferences from less-than-conclusive evidence in the absence of information to the contrary. The examples about birds being able to fly are of this type. If we know that Tweety is a bird, that gives us some evidence that

Tweety can fly, but it is not conclusive. In the absence of information to the contrary, however, we are willing to go ahead and tentatively conclude that Tweety can fly. Now even before we do any detailed analysis of nonmonotonic logic, we can see that there will be problems in interpreting it as a model of default reasoning: In the formal semantics McDermott and Doyle provide for nonmonotonic logic, all the nonmonotonic inferences are valid. Default reasoning, however, is clearly not a form of valid inference.[1]

Consider the belief that lies behind our willingness to infer that Tweety can fly from the fact that Tweety is a bird. It is probably something like most birds can fly, or almost all birds can fly, or a typical bird can fly. To model this kind of reasoning, in a theory whose *only* axioms are "Tweety is a bird" and "Most birds can fly," we ought to be able to infer (nonmonotonically) "Tweety can fly." Now if this were a form of valid inference, we would be guaranteed that the conclusion is true if the premises are true. This is manifestly not the case. The premises of this inference give us a good reason to draw the conclusion, but not the ironclad guarantee that validity demands.

Now reconsider McDermott's formula that yields nonmonotonic inferences about birds being able to fly:

$$\forall x(\mathsf{Bird}(x) \land M(\mathsf{Can\text{-}Fly}(x)) \supset \mathsf{Can\text{-}Fly}(x))$$

McDermott suggests as a gloss of this formula "Most birds can fly," which would indicate that he thinks of the inferences it sanctions as default inferences. But if we read M as "is consistent" as McDermott and Doyle repeatedly tell us to do elsewhere, the formula actually says something quite different: "For all X, if X is a bird and it is consistent to assert that X can fly, then X can fly." Since the inference rule for M is intended to convey "MP is derivable if $\neg P$ is not derivable," the notion of consistency McDermott and Doyle have in mind seems to be that it is consistent to assert P if $\neg P$ is not derivable. McDermott's formula, then, says that the *only* birds that cannot fly are the ones that can be inferred not to fly. If we have a theory whose only axioms are this one and an assertion to the effect that Tweety is a bird, then the conclusion that Tweety can fly *would* be a valid inference. That is, if it is true that Tweety is a bird, and it is true that only birds inferred

[1]In their informal exposition, McDermott and Doyle (1980 p. 44–46) emphasize that their notion of nonmonontonic inference is *not* to be taken as a form of valid inference. If this is the case, their formal semantics cannot be regarded as the "real" semantics of their nonmonotonic logic. At best, it would provide the conditions that *would* have to hold for the inferences to be valid, but this leaves unanswered the question of what formulas of nonmonotonic logic actually *mean*.

not to fly are in fact unable to fly, and Tweety is not inferred not to fly, then it *must* be true that Tweety can fly.

This type of reasoning is not a form of default reasoning at all; it rather seems to be more like reasoning about one's own knowledge or belief. Hence, we will refer to it as *autoepistemic* reasoning. Autoepistemic reasoning, while different from default reasoning, is an important form of commonsense reasoning in its own right. Consider my reason for believing that I do not have an older brother. It is surely not that one of my parents once casually remarked, "You know, you don't have any older brothers," nor have I pieced it together by carefully sifting other evidence. I simply believe that if I did have an older brother I would know about it; therefore, since I don't know of any older brothers, I must not have any. This is quite different from a default inference based on the belief, say, that most MIT graduates are eldest sons, and that, since I am an MIT graduate, I am probably an eldest son.

Default reasoning and autoepistemic reasoning are both nonmonotonic, but for different reasons. Default reasoning is nonmonotonic because, to use a term from philosophy, it is *defeasible*: its conclusions are tentative, so, given better information, they may be withdrawn. Purely autoepistemic reasoning, however, is not defeasible. If you really believe that you already know all the instances of birds that cannot fly, you cannot consistently hold to that belief and at the same time accept new instances of birds that cannot fly.[2]

As Stalnaker (1993) has observed, autoepistemic reasoning is nonmonotonic because the meaning of an autoepistemic statement is context-sensitive; it depends on the theory in which the statement is embedded. If we have a theory whose only two axioms are

$$\mathsf{Bird(Tweety)}$$
$$\forall x(\mathsf{Bird}(x) \wedge M(\mathsf{Can\text{-}Fly}(x)) \supset \mathsf{Can\text{-}Fly}(x)),$$

then MP does not merely mean that P is consistent—it means that P is consistent with the nonmonotonic theory that contains only those two axioms. We would expect Can-Fly(Tweety) to be a theorem of this theory. If we change the theory by adding ¬Can-Fly(Tweety) as an axiom, we then change the meaning of MP to be that P is consistent with the nonmonotonic theory that contains only the axioms

$$\neg\mathsf{Can\text{-}Fly(Tweety)}$$
$$\mathsf{Bird(Tweety)}$$

[2]Of course, autoepistemic reasoning can be combined with default reasoning; we might believe that we know about *most* of the birds that cannot fly. This could lead to defeasible autoepistemic inferences, but their defeasibility would be the result of their also being default inferences.

$$\forall x(\mathsf{Bird}(x) \wedge M(\mathsf{Can\text{-}Fly}(x)) \supset \mathsf{Can\text{-}Fly}(x)),$$

and we would not expect Can-Fly(Tweety) to be a theorem. The operator M changes its meaning with context just as do indexical words in natural language, such as "I," "here," and "now." The nonmonotonicity associated with autoepistemic statements should therefore be no more puzzling than the fact that "I am hungry" can be true when uttered by a particular speaker at a particular time, but false when uttered by a different speaker at the same time or the same speaker at a different time. So we might say that, whereas default reasoning is nonmonotonic because it is defeasible, autoepistemic reasoning is nonmonotonic because it is indexical.

6.3 The Formalization of Autoepistemic Logic

Rather than try directly to analyze McDermott and Doyle's nonmonotonic logic as a model of autoepistemic reasoning, we will first define a logic that demonstrably does model certain aspects of autoepistemic reasoning and then compare nonmonotonic logic with that. We will call our logic, naturally enough, *autoepistemic logic*. The language will be much like McDermott and Doyle's, an ordinary logical language augmented by autoepistemic modal operators. McDermott and Doyle treat consistency as their fundamental notion, so they take M as the basic modal operator and define its dual L to be $\neg M \neg$. Our logic, however, will be based on the notion of belief, so we will take L to mean "is believed," treat it as primitive, and define M as $\neg L \neg$. In any case, this gives us the same notion of consistency as theirs: a formula is consistent if its negation is not believed. Since there are some problems with regard to the meaning of quantifying into the scope of an autoepistemic operator that are not relevant to the main point of this chapter, we will limit our attention to propositional autoepistemic logic.

Autoepistemic logic is intended to model the beliefs of an agent reflecting upon his own beliefs. The primary objects of interest are sets of autoepistemic logic formulas that are interpreted as the total beliefs of such agents. We will call such a set of formulas an *autoepistemic theory*. The truth of an agent's beliefs, expressed as a propositional autoepistemic theory, will be determined by (1) which propositional constants are true in the external world and (2) which formulas the agent believes. A formula of the form LP will be true with respect to an agent if and only if P is in his set of beliefs. To formalize this, we define notions of interpretation and model as follows:

We proceed in two stages. First we define a *propositional interpre-*

tation of an autoepistemic theory T to be an assignment of truth-values to the formulas of the language of T that is consistent with the usual truth recursion for propositional logic and with any arbitrary assignment of truth-values to propositional constants and formulas of the form LP. A *propositional model* of an autoepistemic theory T is a propositional interpretation of T in which all the formulas of T are true. The propositional interpretations and models of an autoepistemic theory are, therefore, precisely those we would get in ordinary propositional logic by treating all formulas of the form LP as propositional constants. We therefore inherit the soundness and completeness theorems of propositional logic; i.e., a formula P is true in all the propositional models of an autoepistemic theory T if and only if it is a tautological consequence of T (i.e., derivable from T by the usual rules of propositional logic).

Next we define an *autoepistemic interpretation* of an autoepistemic theory T to be a propositional interpretation of T in which, for every formula P, LP is true if and only if P is in T. It should be noted that the theory T itself completely determines the truth of any formula of the form LP in all the autoepistemic interpretations of T, independently of the truth assignment to the propositional constants. Hence, for every truth assignment to the propositional constants of T, there is exactly one corresponding autoepistemic interpretation of T. Finally, an *autoepistemic model* of T is an autoepistemic interpretation of T in which all the formulas of T are true. So the autoepistemic interpretations and models of T are just the propositional interpretations and models of T that conform to the intended meaning of the modal operator L.

This gives us a formal semantics for autoepistemic logic that matches its intuitive interpretation. Suppose that the beliefs of an agent situated in a particular world are characterized by the autoepistemic theory T. The world in question will provide an assignment of truth-values for the propositional constants of T, and any formula of the form LP will be true relative to the agent just in case he believes P. In this way, the agent and the world in which he is situated directly determine an autoepistemic interpretation of T. That interpretation will be an autoepistemic model of T, just in case all the agent's beliefs are true in his world.

Given this semantics for autoepistemic logic, what do we want from a notion of inference for the logic? From an epistemological perspective, the problem of inference is the problem of what set of beliefs (theorems) an ideally rational agent would adopt on the basis of his initial premises (axioms). Since we are trying to model the beliefs of a rational agent, the beliefs should be sound with respect the premises; we want a guarantee that the beliefs are true provided that the premises are true. Moreover, since we assume that the agent is *ideally* rational, the beliefs should be semantically complete; we want them to contain everything that the agent would be semantically justified in concluding from his beliefs and from the knowledge that they are his beliefs. An autoepistemic logic that meets these conditions can be viewed as a competence model of reflection upon one's own beliefs. Like competence models generally, it assumes unbounded resources of time and memory, and is therefore not a plausible model of any finite agent. It is, however, the model upon which the behavior of rational agents ought to converge as their time and memory resources increase.

Formally, we will say an autoepistemic theory T is *sound* with respect to an initial set of premises A if and only if every autoepistemic interpretation of T in which all the formulas of A are true is an autoepistemic model of T. This notion of soundness is the weakest condition that guarantees that all of the agent's beliefs are true whenever all his premises are true. Let I be the autoepistemic interpretation of T that is determined by what is true in the actual world (including what the agent actually believes). If all the formulas of T are true in every autoepistemic interpretation of T in which all the formulas of A are true, then all the formulas of T will be true in I if all the formulas of A are true in I; hence, all the agent's beliefs will be true in the world if all the agent's premises are true in the world. However, if there is an autoepistemic interpretation of T in which all the formulas of A are true but some formulas of T are false, then it is possible that I is that interpretation, and that all the agent's premises will be true in the world, but some of his beliefs will not.

Our formal notion of completeness is that an autoepistemic theory T is *semantically complete* if and only if T contains every formula that is true in every autoepistemic model of T. If a formula P is true in every autoepistemic model of an agent's beliefs, then it must be true if all the agent's beliefs are true, and an ideally rational agent should be able to recognize that and infer P. On the other hand, if P is false in some autoepistemic model of the agent's beliefs, then that model, for all he can tell, might be the way the world actually is; he is therefore justified in not believing P.

The next problem is to give a syntactic characterization of the autoepistemic theories that satisfy these conditions. With a monotonic logic, the usual procedure is to define a collection of inference rules to apply to the axioms. For a nonmonotonic logic this is a nontrivial matter. Much of the technical ingenuity of McDermott and Doyle's systems lies simply in their formulation of a coherent notion of nonmonotonic derivability. The problem is that nonmonotonic inference rules do not yield a simple iterative notion of derivability the way monotonic inference rules do. We can view a monotonic inference process as applying the inference rules in all possible ways to the axioms, generating additional formulas to which the inference rules are applied in all possible ways, and so forth. Since monotonic inference rules *are* monotonic, once a formula has been generated at a given stage, it remains in the generated set of formulas at every subsequent stage. Thus the theorems of a theory in a monotonic system can be defined simply as all the formulas that are generated at any stage. The problem with attempting to follow this pattern with nonmonotonic inference rules is that we cannot draw nonmonotonic inferences reliably at any particular stage, since something inferred at a later stage may invalidate them. Lacking such an iterative structure, nonmonotonic systems often use nonconstructive "fixed point" definitions, which do not directly yield algorithms for enumerating the "derivable" formulas, but do define sets of formulas that respect the intent of the nonmonotonic inference rules (e.g., in McDermott and Doyle's fixed points, MP is included whenever $\neg P$ is not included.)

For our logic, it is easiest to proceed by first specifying the closure conditions that we would expect the beliefs of an ideally rational agent to possess. Viewed informally, the beliefs should include whatever the agent could infer either by ordinary logic or by reflecting on what he believes. Stalnaker (1993) has put this formally by suggesting that a set of formulas T that represents the beliefs of an ideally rational agent should satisfy the following conditions:

1. If $P_1, \ldots, P_n$ are in T, and $P_1, \ldots, P_n \vdash Q$, then Q is in T (where "$\vdash$" means ordinary tautological consequence).
2. If P is in T, then LP is in T.
3. If P is not in T, then $\neg LP$ is in T.

Stalnaker (1993, p. 187) describes the state of belief characterized by such a theory as *stable* "in the sense that no further conclusions could be drawn by an ideally rational agent in such a state." We will therefore describe the theories themselves as *stable autoepistemic theories*.

There are a number of interesting observations we can make about stable autoepistemic theories. First we note that, if a stable autoepistemic theory T is consistent, it will satisfy two more intuitively sound conditions:

4. If LP is in T, then P is in T.

5. If $\neg LP$ is in T, then P is not in T.

Condition 4 holds because, if LP were in T and P were not, $\neg LP$ would be in T (by Condition 3) and T would be inconsistent.[3] Condition 5 holds because, if $\neg LP$ and P were both in T, LP would be in T (by Condition 2) and T would be inconsistent.

Conditions 2–5 imply that any consistent stable autoepistemic theory will be both sound and semantically complete with respect to formulas of the form LP and $\neg LP$: If T is such a theory, then LP will be in T if and only if P is in T, and $\neg LP$ will be in T if and only if P is not in T. Thus, all the propositional models of a stable autoepistemic theory are autoepistemic models. Stability implies a soundness result even stronger than this, however. We can show that the truth of any formula of a stable autoepistemic theory depends only on the truth of the formulas of the theory that contain no autoepistemic operators. (We will call these formulas "objective.")

Theorem 1 *If T is a stable autoepistemic theory, then any autoepistemic interpretation of T that is a propositional model of the objective formulas of T is an autoepistemic model of T.*

(The proofs of all theorems are given in the appendix.)

In other words, if all the objective formulas in a stable autoepistemic theory are true, then all the formulas in that theory are true. Given that the objective formulas of a stable autoepistemic theory determine whether the theory is true, it is not surprising that they also determine what all the formulas of the theory are.

Theorem 2 *If two stable autoepistemic theories contain the same objective formulas, then they contain exactly the same formulas.*[4]

Finally, with these characterization theorems, we can prove that the syntactic property of stability is equivalent the semantic property of completeness.

Theorem 3 *An autoepistemic theory T is semantically complete if and only if T is stable.*

By Theorem 3, we know that stability of an agent's beliefs guar-

---

[3]Condition 4 will, of course, also be satisfied by an inconsistent stable autoepistemic theory, since such a theory would include all formulas of autoepistemic logic.

[4]This theorem implies that our autoepistemic logic does not contain any "non-grounded" self-referential formulas, such as one finds in what are usually called "syntactical" treatments of belief. If, instead of a belief operator, we had a belief predicate, Bel, there might be a term p that denotes the formula Bel(p). Whether Bel(p) is believed or not is clearly independent of any objective beliefs. The lack of such formulas constitutes a characteristic difference between sentence-operator and predicate treatments of propositional attitudes and modalities.

antees that they are semantically complete, but stability alone does not tell us whether they are sound with respect to his initial premises. That is because the stability conditions say nothing about what an agent should *not* believe. They leave open the possibility of an agent's believing propositions that are not in any way grounded in his initial premises. What we need to add is a constraint specifying that the only propositions the agent believes are his initial premises and those required by the stability conditions. To satisfy the stability conditions and include a set of premises A, an autoepistemic theory T must include all the tautological consequences of $A \cup \{LP \mid P \text{ is in } T\} \cup \{\neg LP \mid P \text{ is not in } T\}$. Conversely, we will say that an autoepistemic theory T is *grounded* in a set of premises A if and only if every formula of T is included in the tautological consequences of $A \cup \{LP \mid P \text{ is in } T\} \cup \{\neg LP \mid P \text{ is not in } T\}$. The following theorem shows that this syntactic constraint on T and A captures the semantic notion of soundness.

Theorem 4 *An autoepistemic theory T is sound with respect to an initial set of premises A if and only if T is grounded in A.*

From Theorems 3 and 4, we can see that the possible sets of beliefs that an ideally rational agent might hold, given A as his premises, ought to be just the extensions of A that are grounded in A and stable. We will call these the ***stable expansions*** of A. Note that we say "sets", because there may be more than one stable expansion of a given set of premises. For example, consider $\{\neg LP \supset Q, \neg LQ \supset P\}$ as an initial set of premises.[5] The first formula asserts that, if P is not believed, then Q is true; the second asserts that, if Q is not believed, then P is true. In any stable autoepistemic theory that includes these premises, if P is not in the theory, Q will be, and vice versa. But if the theory is grounded in these premises, if P is in the theory there will be no basis for including Q, and vice versa. Consequently, a stable expansion of $\{\neg LP \supset Q, \neg LQ \supset P\}$ will contain either P or Q, but not both.

It can also happen that there are *no* stable expansions of a given set of premises. Consider, for instance, $\{\neg LP \supset P\}$.[6] If T is a stable autoepistemic theory that contains $\neg LP \supset P$, it must also contain P. If P were not in T, $\neg LP$ would have to be in the T, but then P would be in T—a contradiction. On the other hand, if P is in T, then T is not grounded in $\{\neg LP \supset P\}$. Therefore no stable autoepistemic theory can be grounded in $\{\neg LP \supset P\}$.

This seemingly strange behavior results from the indexicality of the

[5] McDermott and Doyle (1980, p. 51) present this example as $\{MC \supset \neg D, MD \supset \neg C\}$.

[6] McDermott and Doyle (1980, p. 51) present this example as $\{MC \supset \neg C\}$.

autoepistemic operator L. Since L is interpreted relative to an entire set of beliefs, its interpretation will change with the various ways of completing a set of beliefs. In each acceptable completion of a set of beliefs, the interpretation of L will change to make that set stable and grounded in the premises. Sometimes, though, no matter how we try to form a complete a set of beliefs, the result never coincides with the interpretation of L in a way that gives us a stable set of beliefs grounded in the premises.

This raises the question of how to view autoepistemic logic *as* a logic. If we consider a set of premises A as axioms, what do we consider the theorems of A to be? If there is a unique stable expansion of A, it seems clear that we want this expansion to be the set of theorems of A. But what if there are several stable expansions of A—or none at all? If we take the point of view of the agent, we have to say that there can be alternative sets of theorems, or no set of theorems of A. This may be a strange property for a logic to possess, but, given our semantics, it is clear why this happens. An alternative (adopted by McDermott and Doyle with regard to their fixed points) is to take the theorems of A to be the intersection of the set of all formulas of the language with all the stable expansions of A. This yields the formulas that are in all stable expansions of A if there is more than one, and it makes the theory inconsistent if there is no stable expansion of A. This too is reasonable, but it has a different interpretation. It represents what an outside observer would know, given only knowledge of the agent's premises and that he is ideally rational.

6.4 Analysis of Nonmonotonic Logic

Now we are in a position to provide an analysis of nonmonotonic logic that will explain its peculiarities in terms of autoepistemic logic. Briefly, our conclusions will be that the original nonmonotonic logic of McDermott and Doyle (1980) is simply too weak to capture the notions they wanted, and that McDermott's (1982) attempt to strengthen the logic does so in the wrong way.

McDermott and Doyle's first logic is very similar to our autoepistemic logic with one glaring exception; its specification includes nothing corresponding to our Condition 2 (if P is in T, then LP is in T). McDermott and Doyle define the nonmonotonic *fixed points* of a set of premises A, corresponding to our stable expansions of A. In the propositional case, their definition is equivalent to the following:

> T is a fixed point of A just in case T is the set of tautological consequences of $A \cup \{\neg LP \mid P \text{ is not in } T\}$.

Our definition of a stable expansion of A, on the other hand, could be stated as

> T is a stable expansion of A just in case T is the set of tautological consequences of $A \cup \{LP \mid P \text{ is in } T\} \cup \{\neg LP \mid P \text{ is not in } T\}$.

In nonmonotonic logic, $\{LP \mid P \text{ is in } T\}$ is missing from the "base" of the fixed points. This makes it possible for there to be nonmonotonic theories with fixed points that contain P but not LP. So, under an autoepistemic interpretation of L, McDermott and Doyle's agents are omniscient as to what they do not believe, but they may know nothing as to what they do believe.

This explains essentially all the peculiarities of McDermott and Doyle's original logic. For instance, they note (1980, p. 69) that MC does not follow from $M(C \wedge D)$. Changing the modality to L, this is equivalent to $\neg LP$ does not follow from $\neg L(P \vee Q)$. The problem is that, lacking the ability to infer LP from P, nonmonotonic logic permits interpretations of L that are more restricted than simple belief. Suppose we interpret L as "inferable in n or fewer steps" for some particular n. P might be inferable in exactly n steps, and $P \vee Q$ in $n + 1$. According to this $\neg L(P \vee Q)$ would be true and $\neg LP$ would be false. Since this interpretation of L is consistent with McDermott and Doyle's definition of a fixed point, $\neg LP$ does not follow from $\neg L(P \vee Q)$. The other example of this kind noted by McDermott and Doyle is that $\{MC, \neg C\}$ has a consistent fixed point, which amounts to saying simultaneously that P is consistent with everything asserted and that P is false. But this set of premises is equivalent to $\{\neg LP, P\}$, which would have no consistent fixed points if LP were forced to be in every fixed point that contains P.

On the other hand, McDermott and Doyle consider it to be a problem that the set of premises $\{MC \supset D, \neg D\}$ has no fixed point in their logic. Restated in terms of L, this set of premises is equivalent to $\{P \supset LQ, P\}$. Every stable autoepistemic theory containing these premises will also contain Q. (If such a theory is consistent, being closed under tautological consequence, it will contain $\neg LQ$ and, therefore, must contain Q to avoid containing LQ. On the other hand, an inconsistent autoepistemic theory will contain Q because it contains everything.) But Q is not contained in any theory grounded in the premises $\{P \supset LQ, P\}$; it is possible for $P \supset LQ$ and P both to be true with respect to an agent while Q is false. So there is no stable expansion of $\{P \supset LQ, P\}$ in autoepistemic logic; hence, this set of premises cannot be the foundation of an appropriate set of beliefs for an ideally rational agent. Thus, our analysis justifies nonmonotonic logic in this case, contrary to the intuition of McDermott and Doyle.

McDermott and Doyle recognized the weakness of the original formulation of nonmonotonic logic, and McDermott (1982) has gone on to develop a group of theories that are stronger because they are based on modal rather than classical logic. McDermott's nonmonotonic modal theories alter the logic in two ways. First, the definition of fixed point is changed to be equivalent to

> T is a fixed point of A just in case T is the set of *modal* consequences of $A \cup \{\neg LP \mid P \text{ is not in } T\}$,

where "modal consequence" means that $P \vdash LP$ is used as an additional inference rule. Second, McDermott considers only theories that include as premises the axioms of one of the standard modal logics "T," "S4," and "S5."

Merely changing the definition of fixed point brings McDermott's logic much closer to autoepistemic logic. In particular, adding $P \vdash LP$ as an inference rule means that all modal fixed points of A are stable expansions of A. However, adding $P \vdash LP$ as an inference rule, rather than adding $\{LP \mid P \text{ is in } T\}$ to the base of T, has as a consequence that not all stable expansions of A are modal fixed points of A. The difference is that, in autoepistemic logic, if P can be derived from LP, then both can be in a stable expansion of the premises, whereas in McDermott's logic there must be a derivation of P that does not rely on LP. Thus, although in autoepistemic logic there is a stable expansion of $\{LP \supset P\}$ that includes P, in McDermott's logic there is no modal fixed point of $\{LP \supset P\}$ that includes P. It is as if, in autoepistemic logic, one can acquire the belief that P and justify it later by the premise that, if P is believed, then it is true. In nonmonotonic logic, however, the justification of P has to precede belief in LP. This makes the interpretation of L in nonmonotonic modal logic more like "justified belief" than simple belief.

Since we have already shown that autoepistemic logic requires no specific axioms to capture a competence model of autoepistemic reasoning, we might wonder what purpose is served by McDermott's second modification of nonmonotonic logic, the addition of the axioms of various modal logics. The most plausible answer is that, besides behaving in accordance with the principles of autoepistemic logic, an ideally rational agent might well be expected to know what some of those principles are. For instance, the modal logic T has all instances of the schema $L(P \supset Q) \supset (LP \supset LQ)$ as axioms. This says that the agent's beliefs are closed under modus ponens—which is true for an ideally rational agent, so he might as well believe it. S4 adds the schema $LP \supset LLP$, which means that, if the agent believes P, he believes that he believes it (Condition 2). S5 adds the schema $\neg LP \supset L\neg LP$, which means that, if the agent does not believe P, he believes that he does not believe it (Condition 3). Since all these formulas are always true with respect to any ideally rational agent, it seems plausible to expect him to adopt them as premises. Thus, S5 seems to be the most plausible candidate of the nonmonotonic logics as a model of autoepistemic reasoning.

The problem is that all of these logics also contain the schema $LP \supset P$, which means that, if the agent believes P, then P is true—but this is not generally true, even for ideally rational agents.[7] It turns out that $LP \supset P$ will always be contained in any stable autoepistemic theory (that is, ideally rational agents always believe that their beliefs are true), but making it a premise allows beliefs to be grounded that otherwise would not be. As a premise the schema $LP \supset P$ can itself be justification for believing P, while as a "theorem" it must be derived from $\neg LP$, in which case P is not believed, or from P, in which case P must be independently justified, or from some other grounded formulas. In any case, as a premise schema, $LP \supset P$ can sanction *any* belief whatsoever in autoepistemic logic. This is not generally true in modal nonmonotonic logic, as we have also seen, but it is true in nonmonotonic S5. The S5 axiom schema $\neg LP \supset L\neg LP$ embodies enough of the model theory of autoepistemic logic to allow LP to be "self grounding": The schema $\neg LP \supset L\neg LP$ is equivalent to the schema $\neg L\neg LP \supset LP$, which allows LP to be justified by the fact that its negation is not believed. This inference is never in danger of being falsified, but, from this and $LP \supset P$, we obtain an unwarranted justification for believing P.

The collapse of nonmonotonic S5 into monotonic S5 follows immediately. Since $LP \supset P$ can be used to justify belief in any formula at all, there are no formulas that are absent from every fixed point of theories based on nonmonotonic S5. It follows that there are no formulas of the form $\neg LP$ that are contained in every fixed point of theories based on nonmonotonic S5; hence there are no theorems of the form $\neg LP$ in any theory based on nonmonotonic S5. (Recall that the theo-

[7] $LP \supset P$ would be an appropriate axiom schema if the interpretation of LP were "P is known" rather than "P is believed," but *that* notion is not nonmonotonic. An agent cannot, in general, know when he does not know P, because he might believe P—leading him to believe that he knows P—while P is in fact false. Since agents are unable to reflect directly on what they do not know (only on what they do not believe), an autoepistemic logic of knowledge would not be a nonmonotonic logic; rather, the appropriate logic would seem to be monotonic S4.

rems are the intersection of all the fixed points.) Since these formulas are just the ones that would be produced by nonmonotonic inference, nonmonotonic S5 collapses to monotonic S5. In more informal terms, an agent who assumes that he is infallible is liable to believe anything, so an outside observer can conclude nothing about what he does not believe.

The real problem with nonmonotonic S5, then, is not the S5 schema; therefore McDermott's rather unmotivated suggestion to drop back to nonmonotonic S4 (1982, p. 45) is not the answer. The S5 schema merely makes explicit the consequences of adopting $LP \supset P$ as a premise schema that are implicit in the logic's natural semantics. If we want to base nonmonotonic logic on a modal logic, the obvious solution is to drop back, not to S4, but to what Stalnaker (1993) calls "weak S5"—S5 without $LP \supset P$. It is much better motivated and, moreover, has the advantage of actually being nonmonotonic.

In autoepistemic logic, however, even this much is unneccessary. Adopting any of the axioms of weak S5 as premises makes no difference to what can be derived. The key fact is the following theorem:

Theorem 5 *If P is true in every autoepistemic interpretation of T, then T is grounded in $A \cup \{P\}$ if and only if T is grounded in A.*

An immediate corollary of this result is that, if P is true in every autoepistemic interpretation of T, then T is a stable expansion of $A \cup \{P\}$ if and only if T is a stable expansion of A.

The modal axiom schemata of weak S5,

$$L(P \supset Q) \supset (LP \supset LQ)$$
$$LP \supset LLP$$
$$\neg LP \supset L\neg LP,$$

simply state Conditions 1–3, so all their instances are true in every autoepistemic interpretation of any stable autoepistemic theory. The nonmodal axioms of weak S5 are just the tautologies of propositional logic, so they are true in every interpretation (autoepistemic or otherwise) of any autoepistemic theory (stable or otherwise). It immediately follows by Theorem 5, therefore, that a set of premises containing any of the axioms of weak S5 will have exactly the same stable expansions as the corresponding set of premises without any weak-S5 axioms.

6.5 Conclusion

McDermott and Doyle recognized that their original nonmonotonic logic was too weak; when McDermott tried to strengthen it, however, he misdiagnosed the problem. Because he was thinking of nonmonotonic logic as a logic of provability rather than belief, he apparently thought the problem was the lack of any connection between provability and truth. At one point he says "Even though $\neg M \neg P$ (abbreviated LP) might plausibly be expected to mean 'P is provable,' there was not actually any relation between the truth values of P and LP," (1982, p. 34), and later he acknowledges the questionability of the schema $LP \supset P$, but says that "it is difficult to visualize any other way of relating provability and truth," (1982, p. 35). If one interprets nonmonotonic logic as a logic of belief, however, there is no reason to expect any connection between the truth of LP and the truth of P. And, as we have seen, the real problem with the original nonmonotonic logic was that the "if" half of the semantic definition of L—that LP is true if and only if P is believed—was not expressed in the logic.

Appendix: Proofs of Theorems

Theorem 1 *If T is a stable autoepistemic theory, then any autoepistemic interpretation of T that is a propositional model of the objective formulas of T is an autoepistemic model of T.*

Proof. Suppose that T is a stable autoepistemic theory and I is an autoepistemic interpretation of T that is a propositional model of the objective formulas of T. All the objective formulas of T are true in I. T must be consistent because an inconsistent stable autoepistemic theory would contain all formulas of the language, which would include many objective formulas that are not true in I. Let P be an arbitrary formula in T. Since stable autoepistemic theories are closed under tautological consequence, T must also contain a set of formulas $P_1, \ldots, P_k$ that taken together entail P, where, for each i between 1 and k, there exist n and m such that P_i is of the form

$$P_{i,1} \vee LP_{i,2} \vee \ldots \vee LP_{i,n} \vee \neg LP_{i,n+1} \vee \ldots \vee \neg LP_{i,m}$$

and $P_{i,1}$ is an objective formula. (Any formula is interderivable with a set of such formulas by propositional logic alone.) There are two cases to be considered:

(1) Suppose at least one of $LP_{i,2}, \ldots, LP_{i,n}, \neg LP_{i,n+1}, \ldots, \neg LP_{i,m}$ is in T. By Conditions 4 and 5, we know that, if any such formula is in T, it must be true in I, since T is consistent and I is an autoepistemic interpretation of T. But, since each of these formulas entails P_i, it follows that P_i is also true in I.

(2) Suppose the first case does not hold. Conditions 2 and 3 guarantee that in every stable autoepistemic theory, for every formula P, either LP or $\neg LP$ will be in the theory. Hence, if T does not contain any of $LP_{i,2}, \ldots, LP_{i,n}, \neg LP_{i,n+1}, \ldots, \neg LP_{i,m}$, it must contain all of

$\neg LP_{i,2}, \ldots, \neg LP_{i,n}, LP_{i,n+1}, \ldots, LP_{i,m}$. But $P_{i,1}$ is a tautological consequence of P_i and these formulas (imagine repeated applications of the resolution principle); so $P_{i,1}$ must be in T. But $P_{i,1}$ is objective, and so, by hypothesis, must be true in I. Since $P_{i,1}$ entails P_i, it must be the case that P_i is true in I.

In either case, P_i will be true in I. All the P_i taken together entail P, so P must also be true in I. Since P was chosen arbitrarily, every formula of T must be true in I; hence I is an autoepistemic model of T. □

Theorem 2 *If two stable autoepistemic theories contain the same objective formulas, then they contain exactly the same formulas.*

Proof. Suppose that T_1 and T_2 contain the same objective formulas and T_1 contains P. We prove by induction on the depth of nesting of autoepistemic operators in P (the "L-depth" of P) that T_2 also contains P. If the L-depth of P is 0, the theorem is trivially true, since P will be an objective formula. Now suppose that P has an L-depth of d greater than 0, and that, if two stable autoepistemic theories contain the same objective formulas, then they contain exactly the same formulas whose L-depth is less than d.

Since stable autoepistemic theories are closed under tautological consequence, T_1 must also contain a set of formulas $P_1, \ldots, P_k$ that are interderivable with P by propositional logic, where, for each i between 1 and k, there exist n and m such that P_i is of the form

$$P_{i,1} \lor LP_{i,2} \lor \ldots \lor LP_{i,n} \lor \neg LP_{i,n+1} \lor \ldots \lor \neg LP_{i,m}$$

and $P_{i,1}$ is an objective formula. Note that, since propositional logic will treat all the formulas of the form $LP_{i,j}$ as propositional constants, it is impossible to increase the L-depth of a formula by propositional inference, so each of these formulas will have an L-depth of not more than d.

We can also assume that T_1 and T_2 are consistent. If one of these theories were inconsistent, it would contain all formulas of the language. Since, by hypothesis, the two theories contain the same objective formulas, the other theory would contain all the objective formulas of the language and, since these formulas are inconsistent, it would also contain all the formulas of the language. For each P_i, there are three cases to be considered:

(1) T_1 contains $LP_{i,j}$ for some j between 2 and n. Since T_1 is consistent, by Condition 4 it must also contain $P_{i,j}$. Since the L-depth of $P_{i,j}$ is one less than that of $LP_{i,j}$, it must be less than d; so, by

hypothesis, T_2 must contain P_{ij} and, by Condition 2, it must contain $LP_{i,j}$. But P_i is a tautological consequence of $LP_{i,j}$, so T_2 contains P_i.

(2) T_1 contains $\neg LP_{i,j}$ for some j between $n+1$ and m. Since T_1 is consistent, by Condition 5 it must not contain $P_{i,j}$. Since the L-depth of $P_{i,j}$ is one less than that of $\neg LP_{i,j}$, it must be less than d; therefore, by hypothesis, T_2 must not contain $P_{i,j}$ and, by Condition 3, it must contain $\neg LP_{i,j}$. But P_i is a tautological consequence of $\neg LP_{i,j}$, so T_2 contains P_i.

(3) Suppose neither of the first two cases holds. Conditions 2 and 3 guarantee that in every stable autoepistemic theory, for every formula P, either LP or $\neg LP$ will be in the theory. Hence, if T_1 does not contain any of $LP_{i,2}, \ldots, LP_{i,n}, \neg LP_{i,n+1}, \ldots, \neg LP_{i,m}$, it must contain all of $\neg LP_{i,2}, \ldots, \neg LP_{i,n}, LP_{i,n+1}, \ldots, LP_{i,m}$. But $P_{i,1}$ is a tautological consequence of P_i and these formulas; so $P_{i,1}$ must be in T_1. $P_{i,1}$ is objective, however, so $P_{i,1}$ must also be in T_2. Since P_i is a tautological consequence of $P_{i,1}$, T_2 contains P_i.

Thus, all of $P_1, \ldots, P_k$ are in T_2. Since P is a tautological consequence of these formulas, P is also in T_2. Since P was chosen arbitrarily, every formula in T_1 is also in T_2. The same argument can be used to show that every formula in T_2 is also in T_1, so T_1 and T_2 contain exactly the same formulas. □

Theorem 3 *An autoepistemic theory T is semantically complete if and only if T is stable.*

Proof. "If" direction: we show that, if T is a stable autoepistemic theory, then T contains every formula that is true in every autoepistemic model of T. Let T be a stable autoepistemic theory and let P be an arbitrary formula that is not in T. We show that there is an autoepistemic model of T in which P is false.

We know from propositional logic that P is propositionally equivalent to (i.e., true in the same propositional models as) the conjunction of a set of formulas $P_1, \ldots, P_k$, where, for each i between 1 and k, there exist n and m such that P_i is of the form

$$P_{i,1} \vee LP_{i,2} \vee \ldots \vee LP_{i,n} \vee \neg LP_{i,n+1} \vee \ldots \vee \neg LP_{i,m}$$

and $P_{i,1}$ is an objective formula. Since P will be a tautological consequence of $P_1, \ldots, P_k$ and T is stable, Condition 1 guarantees that, if P is not in T, at least one of $P_1, \ldots, P_k$ must not be in T. Let P_i be such a formula. P_i is a tautological consequence of each of its disjuncts, so none of them can be in T. We show that there is an autoepistemic model of T in which all of these disjuncts are false.

Since $P_{i,1}$ is not in T, it must not be a tautological consequence of the objective formulas of T. Given this and the fact that $P_{i,1}$ is objective, it follows from the completeness theorem for propositional logic that there must be a truth assignment to the propositional constants of T in which $P_{i,1}$ is false and all the objective formulas of T are true. But, we can extend this truth assignment (or any truth assignment to the propositional constants of T—see Section 6.3) to an autoepistemic interpretation of T. Call this interpretation I and note that $P_{i,1}$ is false in I. I will be a propositional model of the objective formulas of T; so, by Theorem 1, I is an autoepistemic model of T in which $P_{i,1}$ is false.

Now consider the other disjuncts of P_i. Note that Conditions 2 and 3 require that a stable theory contain all the formulas of the form LP or $\neg LP$ that are true in the autoepistemic interpretations of the theory. Since none of $LP_{i,2}, \ldots, LP_{i,n}, \neg LP_{i,n+1}, \ldots, \neg LP_{i,m}$ are in T, none of $LP_{i,2}, \ldots, LP_{i,n}, \neg LP_{i,n+1}, \ldots, \neg LP_{i,m}$ are true in any autoepistemic interpretation of T. In particular, none of $LP_{i,2}, \ldots, LP_{i,n}, \neg LP_{i,n+1}, \ldots, \neg LP_{i,m}$ are true in I. Therefore, I is an autoepistemic model of T in which, since all of the disjuncts of P_i are false, P_i itself is false. But P is propositionally equivalent to a conjunction that includes P_i, so I is an autoepistemic model of T in which P is false.

"Only if" direction: we show that, if T is semantically complete, then T is stable. Suppose T is semantically complete. For any formula P, if P is true in every autoepistemic model of T, then P is in T. Let I be an arbitrary autoepistemic model of T. If we can show that some formula P is true in I, P must be true in every autoepistemic model of T (since I is arbitrarily chosen) and, thus, P must be in T. We now show that T satisfies Conditions 1–3.

(1) Suppose $P_1, \ldots, P_n$ are in T and $P_1, \ldots, P_n \vdash Q$. Since I is a model of T, $P_1, \ldots, P_n$ will be true in I. Since $P_1, \ldots, P_n$ will is true in I and Q is a tautological consequence of $P_1, \ldots, P_n$, Q will also be true in I. Therefore, Q will be in T. (2) Suppose P is in T. Since I is an autoepistemic model of T, LP will be true in I. Therefore, LP will be in T. (3) Suppose P is not in T. Since I is an autoepistemic model of T, $\neg LP$ will be true in I. Therefore, $\neg LP$ will be in T.

Conditions 1–3 are all satisfied, so T is stable. □

Theorem 4 *An autoepistemic theory T is sound with respect to an initial set of premises A if and only if T is grounded in A.*

Proof. "If" direction: suppose T is grounded in A. Every formula of T is therefore included in the tautological consequences of $A \cup \{LP \mid P \text{ is in } T\} \cup \{\neg LP \mid P \text{ is not in } T\}$. We show that T is sound with respect to A—i.e., that every autoepistemic interpretation of T in which all the formulas of A are true is an autoepistemic model of T.

Let I be an autoepistemic interpretation of T in which all the formulas in A are true. We show that I is an autoepistemic model of T. If P is in A, then, trivially, P is true in I. If P is of the form LQ and Q is in T, or if P is of the form $\neg LQ$ and Q is not in T, then P is true in I because I is an autoepistemic interpretation of T. We have now shown that all the formulas in $A \cup \{LP \mid P \text{ is in } T\} \cup \{\neg LP \mid P \text{ is not in } T\}$ are true in I, so all their tautological consequences are true in I. But all the formulas of T are included in this set, so I is an autoepistemic model of T. Since I was an arbitrarily chosen autoepistemic interpretation of T in which all the formulas of A are true, every autoepistemic interpretation of T in which all the formulas of A are true is an autoepistemic model of T.

"Only if" direction: suppose T is sound with respect to A. Every autoepistemic interpretation of T in which all the formulas of A are true is therefore an autoepistemic model of T. We show that T is grounded in A—i.e., every formula of T is a tautological consequence of $A \cup \{LP \mid P \text{ is in } T\} \cup \{\neg LP \mid P \text{ is not in } T\}$.

Let $A' = A \cup \{LP \mid P \text{ is in } T\} \cup \{\neg LP \mid P \text{ is not in } T\}$. Note that, for all P, if P is in T, LP will be in A', so LP will be true in every propositional model of A'; however, if P is not in T, $\neg LP$ will be in A' and LP will not be true in any propositional model of A'. Therefore, in every propositional model of A', LP is true if and only if P is in T, so every propositional model of A' is an autoepistemic interpretation of T. Since every autoepistemic interpretation of T in which all the formulas of A are true is an autoepistemic model of T, every propositional model of A' is an autoepistemic model of T. Since every formula in T is true in in every autoepistemic model of T, every formula in T is true in every propositional model of A'. By the completeness theorem for propositional logic, every formula of T is therefore a tautological consequence of A'. Hence T is grounded in A. □

Theorem 5 *If P is true in every autoepistemic interpretation of T, then T is grounded in $A \cup \{P\}$ if and only if T is grounded in A.*

Proof. Suppose that P is true in every autoepistemic interpretation of T. For any set of premises A, the set of autoepistemic interpretations of T in which every formula of $A \cup \{P\}$ is true is therefore the same as the set of autoepistemic interpretations of T in which every formula of A is true. Thus, every autoepistemic interpretation of T in which every formula of $A \cup \{P\}$ is true is an autoepistemic model of T if and only if every autoepistemic interpretation of T in which every formula of A is true is an autoepistemic model of T. Hence, T is sound with respect to $A \cup \{P\}$ if and only if T is sound with respect to A. By Theorem 4, therefore, T is grounded in $A \cup \{P\}$ if and only if T is grounded in A. □

7

Possible-World Semantics for Autoepistemic Logic

7.1 Introduction

In Chapter 6, we presented a nonmonotonic logic for modeling the beliefs of ideally rational agents who reflect on their own beliefs, which we called "autoepistemic logic." We defined a simple and intuitive semantics for autoepistemic logic and proved the logic sound and complete with respect to that semantics. However, the nonconstructive character of both the logic and its semantics made it difficult to prove the existence of sets of beliefs satisfying all the constraints of autoepistemic logic. This note presents an alternative, possible-world semantics for autoepistemic logic that enables us to construct finite models for autoepistemic theories, as well as to demonstrate the existence of sound and complete autoepistemic theories based on given sets of premises.

Autoepistemic logic is nonmonotonic, because we can make statements in the logic that allow an agent to draw conclusions about the world from his own lack of information. For example, we can express the belief that "If I do not believe P, then Q is true." If an agent adopts this belief as a premise and he has no means of inferring P, he will be able to derive Q. On the other hand, if we add P to his premises, Q will no longer be derivable. Hence, the logic is nonmonotonic.

Autoepistemic logic is closely related to the nonmonotonic logics of McDermott and Doyle (1980; McDermott 1982). In fact, it was designed to be a reconstruction of these logics that avoids some of their

The research reported herein was supported by the Air Force Office of Scientific Research under Contract No. F49620-82-K-0031. The views and conclusions expressed in this document are those of the author and should not be interpreted as necessarily representing the official policies or endorsements, either expressed or implied, of the Air Force Office of Scientific Research or the U. S. Government.

peculiarities. This is discussed in detail in Chapter 6. This work is also closely related to that of Halpern and Moses (1984), the chief difference being that theirs is a logic of knowledge rather than belief. Finally, Levesque (1981) has also developed a kind of autoepistemic logic, but in his system the agent's premises are restricted to a sublanguage that makes no reference to what he believes.

7.2 Summary of Autoepistemic Logic

The language of autoepistemic logic is that of ordinary propositional logic, augmented by a modal operator L. We want formulas of the form LP to receive the intuitive interpretation "P is believed" or "I believe P." For example, $P \supset LP$ could be interpreted as saying "If P is true, then I believe that P is true."

The type of object that is of primary interest in autoepistemic logic is a set of formulas that can be interpreted as a specification of the beliefs of an agent reflecting upon his own beliefs. We will call such a set of formulas an *autoepistemic theory*. The truth of an agent's beliefs, expressed as an autoepistemic theory, is determined by (1) which propositional constants are true in the external world and (2) which formulas are believed by the agent. A formula of the form LP will be true with respect to an agent if and only if P is in his set of beliefs. To formalize this, we define the notions of autoepistemic interpretation and autoepistemic model. An *autoepistemic interpretation* I of an autoepistemic theory T is a truth assignment to the formulas of the language of T that satisfies the following conditions:

1. I conforms to the usual truth recursion for propositional logic.
2. A formula LP is true in I if and only if $P \in T$.

An *autoepistemic model* of T is an autoepistemic interpretation of T in which all the formulas of T are true. (Any truth assignment satisfying Condition 1 in which all the formulas of T are true will be called simply a *model* of T.)

We can readily define notions of soundness and completeness relative to this semantics. Soundness of a theory must be defined with respect to some set of premises. Intuitively speaking, an autoepistemic theory T, viewed as a set of beliefs, will be sound with respect to a set of premises A, just in case every formula in T must be true, given that all the formulas in A are true and given that T is, in fact, the set of beliefs under consideration. This is expressed formally by the following definition:

An autoepistemic theory T is *sound* with respect to a set of premises

A if and only if every autoepistemic interpretation of T that is a model of A is also a model of T.

The definition of completeness is equally simple. A semantically complete set of beliefs will be one that contains everything that must be true, given that the entire set of beliefs is true and given that it is the set of beliefs being reasoned about. Stated formally, this becomes

An autoepistemic theory T is *semantically complete* if and only if T contains every formula that is true in every autoepistemic model of T.

Finally, we can give syntactic characterizations of the autoepistemic theories that conform to these definitions of soundness and completeness (Chapter 6, Theorems 3 and 4). We say that an autoepistemic theory T is *stable* if and only if (1) it is closed under ordinary tautological consequence, (2) $LP \in T$ whenever $P \in T$, and (3) $\neg LP \in T$ whenever $P \notin T$.

Theorem 1 *An autoepistemic theory T is semantically complete if and only if T is stable.*

We say that an autoepistemic theory T is *grounded* in a set of premises A if and only if every formula in T is a tautological consequence of $A \cup \{LP \mid P \in T\} \cup \{\neg LP \mid P \notin T\}$.

Theorem 2 *An autoepistemic theory T is sound with respect to a set of premises A if and only if T is grounded in A.*

With these soundness and completeness theorems, we can see that the possible sets of beliefs an ideally rational agent might hold, given A as his premises, would be stable autoepistemic theories that contain A and are grounded in A. We call these theories *stable expansions* of A.

7.3 An Alternative Semantics for Autoepistemic Logic

The semantics we have provided for autoepistemic logic is simple, intuitive, and allows us to prove a number of important general results, but it requires enumerating an infinite truth assignment if the theory under consideration contains infinitely many formulas. This makes it difficult to exhibit particular models and interpretations we may be interested in. The problem is that, in the general case, there need be no systematic connection between the truth of one formula of the form LP and any other. Autoepistemic logic is designed to characterize the beliefs of ideally rational agents, but we want the semantics to be broader than that. The semantics we have defined is intended to apply to arbitrary sets of beliefs, with the beliefs of ideally rational agents being a special case (just as model theory for standard logic applies to arbitrary sets of formulas, not just to those that are closed under logical consequence). Thus, our semantics makes no necessary connection between the truth of $L(P \wedge Q)$ and LP or LQ, because it is at least conceivable that an agent might be so logically deficient as to believe $P \wedge Q$ without believing P or believing Q. In such a case, there is little we can expect the truth definition for an autoepistemic theory to do, other than to list the true formulas of the form LP by brute stipulation.

If we confine our attention to ideally rational agents, however, much more structure emerges. In fact, we can show that stable autoepistemic theories can be simply characterized by Kripke-style possible-world models for modal logic (1971). For our purposes, what we need to recall about a Kripke structure is that it contains a set of possible worlds and an accessibility relation between pairs of worlds. The truth of a formula is defined relative to a world, and conforms to the usual truth recursion for propositional logic. A formula of the form LP is true in a world W just in case P is true in every world accessible from W. Kripke structures in which the accessibility relation is an equivalence relation are called S5 structures, and the S5 structures that will be of interest to us are those in which every world is accessible from every world. We will call these the *complete* S5 structures. Our major result is that the sets of formulas that are true in every world of some complete S5 structure are exactly the stable autoepistemic theories. (This result has been obtained independently by Halpern and Moses (1984) and by Melvin Fitting (personal communication)).

Theorem 3 *T is the set of formulas that are true in every world of some complete S5 structure if and only if T is a stable autoepistemic theory.*

Proof. Suppose T is the set of formulas true in every world of a complete S5 structure. By the soundness of propositional logic, T is closed under tautological consequence. By the truth rule for L, LP is true in every world just in case P is true in every world; therefore $LP \in T$ if and only if $P \in T$. Furthermore, by the truth rule for L, LP is false in every world just in case P is false in some world; so $\neg LP \in T$ if and only if $P \notin T$. Therefore T is stable. In the opposite direction, suppose that T is stable. Let T' be the set of formulas of T that contain no occurrences of L. We will call these the *objective* formulas of T. Since T is closed under tautological consequence, T' will also be closed under tautological consequence. Consider the set of all models of T' and the complete S5 structure in which each of these models defines a possible world. T' will contain exactly the objective formulas true in every world in this model; hence, T' will contain precisely the objective formulas of the stable autoepistemic theory T'' defined by this S5 structure. But by a previous result (Chapter 6, Theorem 2), stable theories containing the same objective formulas are identical, so T must be the same as T''. Hence, T is the set of formulas true in every world of a complete S5 structure. □

Given this result, we can characterize any autoepistemic interpretation of any stable theory by an ordered pair consisting of a complete S5 structure (to specify the agent's beliefs) and a propositional truth assignment (to specify what is actually true in the world). Such a structure (K, V) defines an autoepistemic interpretation of the theory T consisting of all the formulas that are true in every world in K. A formula of T is true in (K, V) if it is true according to the standard truth recursion for propositional logic, where the propositional constants are true in (K, V) if and only if they are true in V, and the formulas of the form LP are true in (K, V) if and only if they are true in every world in K (using the truth rules for Kripke structures). We will say that (K, V) is a *possible-world interpretation* of T and, if every formula of T is true in (K, V), we will say that (K, V) is also a *possible-world model* of T. In view of the preceding theorem, it should be obvious that for every autoepistemic interpretation or autoepistemic model of a stable theory there is a corresponding possible-world interpretation or possible-world model, and vice versa.

Theorem 4 *If (K, V) is a possible-world interpretation of T, then (K, V) will be a possible-world model of T if and only if the truth assignment V is consistent with the truth assignment provided by one of the possible worlds in K (i.e., if the actual world is one of the worlds that are compatible with what the agent believes).*

Proof. If V is compatible with one of the worlds in K, then any propositional constant that is true in all worlds in K will be true in V. Therefore, any formula that comes out true in all worlds in K will also come out true in (K, V), and (K, V) will be a possible-world model of T. In the opposite direction, suppose that V is not compatible with any of the worlds in K. Then, for each world W in K, there will be some propositional constant that W and V disagree on. Take that constant or its negation, whichever is true in W, plus the corresponding formulas for all other worlds in K, and form their disjunction. (This will be a finite disjunction, provided there are only finitely many propositional constants in the language.) This disjunction will be true in every world in K, so it will be a formula of T, but it will be false in V. Therefore, (K, V) will not be a possible-world model of T. □

7.4 Applications of Possible-World Semantics

One of the problems with our original presentation of autoepistemic logic was that, since both the logic and its semantics were defined nonconstructively, we were unable to easily prove the existence of stable expansions of nontrivial sets of premises. With the finite models provided by the possible-world semantics for autoepistemic logic, this becomes quite straightforward. For instance, we claimed (see Chapter 6) that the set of premises $\{\neg LP \supset Q, \neg LQ \supset P\}$ has two stable expansions—one containing P but not Q, and the other containing Q but not P—but we were unable to do more than give a plausibility argument for that assertion. We can now demonstrate this fact quite rigorously.

Consider the stable theory T, generated by the complete S5 structure that contains exactly two worlds, $\{P, Q\}$ and $\{P, \neg Q\}$. (We will represent a possible world by the set of propositional constants and negations of propositional constants that are true in it.) The possible-world interpretations of T will be the ordered pairs consisting of this S5 structure and any propositional truth assignment. Consider all the possible-world interpretations of T in which $\neg LP \supset Q$ and $\neg LQ \supset P$ are both true. By exhaustive enumeration, it is easy to see that these are exactly

$$(\{\{P, Q\}, \{P, \neg Q\}\}, \{P, Q\})$$
$$(\{\{P, Q\}, \{P, \neg Q\}\}, \{P, \neg Q\})$$

Since, in each case, the actual world is one of the worlds that are compatible with everything the agent believes, each of these is a possible-world model of T. Therefore, T is sound with respect to $\{\neg LP \supset Q, \neg LQ \supset P\}$. Since T is stable and includes $\{\neg LP \supset Q, \neg LQ \supset P\}$ (note that both these formulas are true in all worlds in the S5 structure), T is a stable expansion of A. Moreover, it is easy to see that T contains P but not Q. A similar construction yields a stable expansion of T that contains Q but not P.

On the other hand, if both P and Q are to be in a theory T, the corresponding S5 structure contains only one world, $\{P, Q\}$. But then $(\{\{P, Q\}\}, \{\neg P, \neg Q\})$ is a possible-world interpretation of T in which $\neg LP \supset Q$ and $\neg LQ \supset P$ are both true, but some of the formulas of T

are not (P and Q, for instance). Hence, if T contains both P and Q, T is not a stable expansion of $\{\neg LP \supset Q, \neg LQ \supset P\}$.

8

Autoepistemic Logic Revisited

"Semantical Considerations on Nonmonotonic Logic" (see Chapter 6) started off to be a short commentary of a methodological/philosophical character on McDermott and Doyle's work on nonmonotonic logics (McDermott and Doyle, 1980; McDermott, 1982). When I started writing the paper, I didn't understand the technical details of McDermott and Doyle's logics very well, but I knew that they had some peculiar and unintuitive properties, and I believed that these might be related to what I saw as some methodological problems in their approach. The principal problem I saw was in trying to model jumping to conclusions by default with a logic whose notion of inference is guaranteed by its semantics to be truth-preserving. To drive home the point, I tried to distinguish between default reasoning and what I called "autoepistemic reasoning," or reasoning about one's own beliefs. (To be linguistically pure, I should have called it "autodoxastic reasoning," but in all honesty, that just didn't have the same ring to it.) I won't go any further into the details here, because that short methodological commentary survives as part (Section 6.2) of what turned out to be a work of much broader scope. As I studied McDermott and Doyle's papers in more detail, I discovered that the problematical features of their logics had technical remedies that could be motivated within a framework based on autoepistemic reasoning. Therefore, I called my reconstruction of nonmonotonic logic "autoepistemic logic." The result was a simple and elegant logic that both explained and eliminated many of the unintuitive properties of McDermott and Doyle's logics, and that also turned out to provide a foundation for a substantial amount of further work.

This material was first presented at the 1983 International Joint Conference on Artificial Intelligence (although it was not published in

This chapter previously appeared in *Artificial Intelligence*, Vol. 25, Nos. 1–2, 1993, and is reprinted with the permission of the publisher, Elsevier, Amsterdam.

Artificial Intelligence until 1985). In 1984, I presented a second paper (see Chapter 7), providing a possible-world semantics for autoepistemic logic. The original work had been based on a syntactic notion of belief: The beliefs of an agent were characterized simply by an arbitrary list of formulas. The original paper went on to develop a theory of an ideal autoepistemic reasoner, but the basic framework could be applied to any reasoner, ideal or not. By confining its scope to ideal reasoners, the second paper is able to develop a more structured model theory that makes concrete examples much easier to present. In particular, in the original framework, the characterization of an ideal reasoner required an infinite number of formulas in the syntactic model, since an ideal reasoner always has an infinite number of beliefs. With the possible-world framework, however, the beliefs of an ideal reasoner based on simple premises can be characterized by a simple, finite model, which makes it easy to rigorously demonstrate the existence of autoepistemic theories having particular properties.

The publication of these two papers was followed by considerable activity by other researchers. Much of this work attempts to relate autoepistemic logic to other formalisms. This literature has become far too extensive to catalogue here, but some of the more interesting papers include Konolige's (1988) and Marek and Truszczynski's (1989) studies of the relationship between autoepistemic logic and default logic, Gelfond's (1987) and Gelfond and Lifschitz's (1988) work on the relation of autoepistemic logic to negation-as-failure in logic programming, and Przymusinski's (1991) grand unification of three-valued forms of all the major formalisms for nonmonotonic reasoning using the "well-founded semantics" for logic programming. In addition, Shvarts (1990) has gone back to look more closely at the relationship between autoepistemic logic and McDermott's (1982) nonmonotonic modal logic. Shvarts has shown that autoepistemic logic does, in fact, fall within McDermott's framework and would be nonmonotonic K45, or nonmonotonic weak S5, to use the terminology of "Semantical Considerations." McDermott, however, looked only at nonmonotonic T, S4, and S5, and so missed out on the appropriate logic for the autoepistemic interpretation of the modal operators.

To me, the most interesting open problems connected with autoepistemic logic concern its extension from propositional logic to first-order logic and the computational properties of the resulting systems. The original papers on autoepistemic logic concerned only the propositional version of the logic. With the possible-world semantics, it is easy to show that any propositional autoepistemic theory is decidable, as long as there are only finitely many proposition letters. For each such theory, there are only finitely many possible-world models, which are themselves finite structures. So any questions of validity, satisfiability, or consequence can be answered simply by enumerating and checking all the models.

If autoepistemic logic is extended to first-order logic, but "quantifying-in" is dissallowed—that is, if autoepistemic modal operators are never applied to formulas with free variables—then all the important syntactic and semantic properties of the logic seem to carry over, but the computational properties change because the models are no longer guaranteed to be finite, and there may be infinitely many of them. In fact, it is easy to see that an autoepistemic version of an essentially undecidable theory would not even be recursively enumerable. The reason is that formulas of the form $\neg LP$ will mean "P is not provable." So if we could enumerate the formulas of the autoepistemic theory, we would have a way to decide the formulas of (perhaps an extension of) the original theory. But if the original theory is essentially undecidable (e.g., Peano arithmetic), this is known to be impossible.

The interesting question then, is what happens with an autoepistemic version of a *decidable* theory. If no extra axioms are added, then it is easy to show that the theory remains decidable. If a finite number of extra axioms are added, the theory remains decidable, as long as quantifying-in is not allowed. This might seem to give us a lot, but in fact such theories are not very expressive. Without quantifying-in, the only way to express a generalization such as "My only brothers are the ones I know about," is to use an axiom schema, which amounts to adding an infinite number of axioms. If we allow axiom schemata or other infinite sets of axioms, it is an open question whether the theory remains decidable.

Finally, the extension of first-order autoepistemic logic to allow quantifying-in remains unsettled. This is a conceptually difficult area, because it is not completely clear what such formulas mean. There is a long and unresolved debate in the philosophy of language about the difference between "it is believed that something has the property P," and "there is something that is believed to have the property P." Yet that is exactly the distinction that would be marked in autoepistemic logic by the difference between $L\exists x P(x)$ and $\exists x LP(x)$. Levesque (1990) has proposed a logic containing an operator whose intuitive interpretation is meant to be "all that I know," which could be thought of as a metatheory for autoepistemic logic and which does allow quantifying-in. However, Levesque has been unable to prove that his logic is semantically complete. In any case, at this writing, I am unaware of any published attempts to allow quantifying-in directly in autoepistemic logic, although a paper by Konolige (1991) on the subject is in press. It will be interesting to see whether a consensus can be reached on the right approach to this problem.

Part IV

Semantics of Natural Language

9

Events, Situations, and Adverbs

9.1 Introduction

This chapter concerns a dispute about the relationship of sentences to the events they describe, and how that relationship is manifested in sentences with adverbial modifiers. The two sides to the argument might be called the "Davidsonian position" and the "situation semantics position"; the former being chiefly represented by Donald Davidson's well-known paper "The Logical Form of Action Sentences" (Davidson 1967b) and the latter by John Perry's critique of Davidson's view, "Situations in Action" (Perry 1983).[1]

The issue turns on Davidson's analysis of how a sentence such as (1) is related to a similar sentence with an adverbial modifier, such as (2).

(1) Jones buttered the toast.

(2) Jones buttered the toast in the bathroom.

Stated very informally, Davidson's position is this: (1) claims that an event of a certain type took place, to wit, a buttering of toast by Jones, and that (2) makes a similar claim but adds that the event took place in the bathroom. Put this way, an advocate of situation semantics could find little to complain about. Perry and Barwise themselves say rather similar things. The dispute is over the way that (1) and (2) claim that certain events took place. Davidson suggests that the event in question is, in effect, a hidden argument to the verb "butter". As he would put it, the logical form of (1), (not analyzing the tense of the verb or the structure of the noun phrase) is not

This research was supported in part by the Air Force Office of Scientific Research under Contract No. F49620-85-K-0012 and in part by a gift from the System Development Foundation.

[1] This dispute is a special case of a much deeper disagreement about semantics that is treated in depth by Barwise and Perry in *Situations and Attitudes* (1983).

(3) Buttered(Jones, the toast)

but rather

(4) $\exists x$(Buttered(Jones, the toast, x)),

where the variable x in (4) ranges over events. Adding the adverbial modifier is then quite straightforward; it is simply an additional predication of the event:

(5) $\exists x$(Buttered(Jones, the toast, x) $\wedge$ In(the bathroom, x))

Perry objects strenuously to making the event described by the sentence an explicit argument to the relation expressed by the verb. He says:

> If we ask what about the statement tells us that there was an event of that type, the only reasonable answer is that the whole statement does. It is not that part of the statement refers to an event, and the other part tells us what it was like. Part of the statement refers to Jones and the other part tells us what he did. Both parts working together tell us that an event of a certain sort occurred. The simple parts of the sentence refer to basic uniformities across events: Jones, buttering, and the toast. The way the simple parts are put together in the sentence describes the event (Perry 1983, p. 2).

Now it happens that Davidson considers but rejects an analysis derived from Reichenbach (1947, p. 266–274) that is in the spirit of Perry's objection. On this analysis, (1) and (2) would be rendered by (6) and (7), respectively:

(6) $\exists x$(x consists in the fact that Jones buttered the toast)

(7) $\exists x$(x consists in the fact that Jones buttered the toast and
x took place in the bathroom)

This seems to meet Perry's objection in that it is the whole statement "Jones buttered the toast" that gives rise to the reference to the event, rather than a hidden argument to the verb. Davidson rejects the analysis, however, on the grounds that its logical properties are problematical. Davidson notes that, from the identity of the Morning Star and Evening Star, we would want to be able to infer that, if I flew my spaceship to the Morning Star, I flew my spaceship to the Evening Star. On the analysis under consideration, this requires being able to infer (9) from (8).

(8) $\exists x$(x consists in the fact that I flew my spaceship to the
Morning Star)

(9) $\exists x$(x consists in the fact that I flew my spaceship to the Evening Star)

Davidson argues that the only reasonable logical principles that would permit this inference would entail the identity of all actually occuring events, which would be absurd. Barwise and Perry's (1983, p. 24–26) rejoinder to this is that Davidson makes the unwarranted assumption that logically equivalent sentences would have to be taken to describe the same event, an idea they reject. Perry (1983) goes on to develop, within the framework of situation semantics, an analysis of event sentences and adverbial modification that is faithful to the idea that, in general, it is an entire sentence that describes an event.[2]

To summarize the state of the argument: Davidson and Perry agree that sentences describe events, but Davidson thinks that it is virtually incoherent to view the event as being described, as it were, "holistically" by the entire sentence, whereas Perry views it as "the only reasonable answer." Barwise and Perry pinpoint where they think Davidson's argument goes wrong, and Perry provides an analysis of adverbial modification consistent with the holistic view.

9.2 Some Facts about Adverbs and Event Sentences

One of the things that Perry's and Davidson's analyses have in common is that neither is based on a very extensive survey of the linguistic data to be explained by a theory of adverbial modification; their arguments are based more on general logical and metaphysical concerns. A close examination of the relevant linguistic phenomena, however, shows that neither Davidson nor Perry have the story quite right, and that a more complete account of adverbial modification has to include at least two possible relations between sentences and events, one close to Davidson's account and the other close to Perry's.

The key set of data we will try to explain is that there exists a significant class of adverbs that can be used to modify event sentences in two quite distinct ways:

(10) (a) John spoke to Bill rudely.
(b) Rudely, John spoke to Bill.

(11) (a) John stood on his head foolishly.
(b) Foolishly, John stood on his head.

[2]We omit the details of Perry's own analysis of adverbial modification, as it is not really needed for the points we wish to make.

(12) (a) John sang strangely.
(b) Strangely, John sang.

The difference between the first and second member of each pair should be clear. For instance, (10a) suggests that it was the way that John spoke to Bill was rude, while (10b) says that the very fact that John spoke to Bill was rude. Thus (10a) leaves open the possibility that John could have spoken to Bill without being rude, but (10b) does not. Similar remarks apply to the other pairs. With this class of adverbs, in general, "*X* did *Y* *Adj*-ly" means that the way *X* did *Y* was *Adj*, and "*Adj*-ly, *X* did *Y*" means that the fact that *X* did *Y* was *Adj*. We will therefore say that the (a) sentences involve a "manner" use of the adverb and that the (b) sentences involve a "fact" use.

One notable observation about the fact use of these adverbs is that they are indeed "factive" in the sense that the truth of the sentence with the adverb entails the truth of the sentence without the adverb. This is in contrast to other "sentential" adverbs like "allegedly" or "probably":

(13) Probably John likes Mary.
(14) John likes Mary

The truth of (13) would not necessarily imply the truth of (14). This factivity extends to the adjective forms from which the adverbs derive:

(15) It was rude for John to speak to Bill.
(16) It was foolish for John to stand on his head.
(17) It was strange for John to sing a song.

Another significant fact is that with copular constructions, only the fact use is possible; the manner use doesn't exist:

(18) Strangely, John is tall.
(19) *John is tall strangely.

Copular constructions accept the fact use of adverbs, as is shown by (18). If we move the adverb to the end of the sentence to try to obtain a manner interpretation as in (19), the sentence is unacceptable.

Finally, perhaps the most important logical difference between the fact and manner uses of these adverbs is that the manner sentences are extensional with respect to the noun phrases in the sentence, whereas the fact sentences are not. That is, we may freely substitute coreferential singular terms in the manner sentences, but not in the fact sentences. Suppose it is considered rude to speak to the Queen (unless, say, she speaks to you first), and suppose John is seated next to the Queen. Then it could well be that (20) is true, while (21) is false, although they differ only in substituting one singular term for a coreferring one.

(20) Rudely, John spoke to the Queen.

(21) Rudely, John spoke to the woman next to him.

Thus (21) can differ in truth-value from (20) because, on at least one interpretation, it seems to entail that it was rude for John to speak to the woman next to him, whoever she was, i.e., even if she were not the Queen. The issue is somewhat complicated by the fact that these sentences also exhibit the sort of *de dicto/de re* ambiguity common to most nonextensional constructs. That is, (20) and (21) seem to be open to an an additional interpretation that there is a certain woman, whom we may identify either as the Queen or the woman next to John, and that it was rude for John to speak to that particular woman.

On the other hand, it seems that (22) and (23) must have the same truth-value on any interpretation, so long as the Queen and the woman next to John are the same person. Moreover, no *de dicto/de re* distinction seems to obtain.

(22) John spoke to the Queen rudely.

(23) John spoke to the woman next to him rudely.

Note, however, that (22) and (23) are not completely extensional in the sense that first-order logic is extensional. That notion of extensionality requires, not only intersubstitutivity of coreferring singular terms, but also intersubstitutivity of sentences with the same truth-value. But even if (24) and (25) have the same truth-value, it does not follow that (26) and (27) do.

(24) John spoke to the Queen.

(25) John spoke to the Prince.

(26) John spoke to the Queen rudely.

(27) John spoke to the Prince rudely.

This sort of behavior is quite general with these adverbs. Examples similar to (20) through (27) can be constructed for "foolishly", "strangely", and all the other adverbs in this class.

9.3 Situations and Events

Before we can give a semantic analysis of event sentences that accounts for these observations, we must develop the framework within which the analysis will be couched. As this will require technical notions of situation and event, this section is devoted to explaining those concepts.

A word of caution is in order before proceeding further. The goal of this exercise is semantic analysis of natural language, not the discovery of Deep Metaphysical Truths. If we postulate situations or events as entities in the world, it is not necessarily because we believe they objectively exist, but because postulating them gives the most natural analysis of the meanings of the class of sentences we are trying to analyze. Our real concern is to identify the metaphysics embedded in the language, not to decide whether that metaphysics is true.

A second word of warning concerns our use of the term "situation". This term is so closely identified with the work of Barwise and Perry, that one might be misled into assuming that the theory of situations assumed here is simply Barwise and Perry's theory. That is emphatically not the case. Yet it seems so clear that both Barwise and Perry's theory and the theory presented here are attempts to formalize a single intuitive notion, that in the end it would probably be even more misleading to employ a different term.

Situations and Propositions

Relatively little in the way of a theory of situations is actually needed to construct an analysis of the linguistic data that we have presented. We really need to say little more than (1) that situations are part of the causal order of the world rather than an abstraction of it, and (2) that situations are in one-to-one correspondence with true propositions. To leave the theory of situations at this, however, would leave open so many questions about what sort of objects situations and propositions were that it might cast serious doubt over the application of the theory to the analysis of event sentences.

In our theory, situations are simpler entities than in Barwise and Perry's theory. For us, a situation is a piece of reality that consists of an n-tuple of entities having an n-ary property or relation.[3] Like Barwise and Perry, then, we take properties to be first-class entities. A proposition is simply an abstraction of a situation: a way that a situation could be. We will assume that for every n-ary property and every n-tuple of entities, there exists the proposition that those entities satisfy that property. That is, suppose we have an individual John and the property of being tall. If John is tall, then there is an actual situation of John being tall. Even if John is not tall, however, there is the abstract possibility of John being tall: i.e., there might have been a John-being-tall situation, but as things turned out, there was not.

[3]We might want to add "at a spatio-temporal location", but we will ignore this aspect of the problem, as the issue seems independent of the others considered here.

This abstract possibility is what we take a proposition to be. A true proposition is one that is the abstraction of an actual situation. We can ask what would be the individuation criteria for situations and for propositions in this theory, and while various answers are possible, the most natural one would be that identity of the properties and each pair of corresponding arguments are required for the identity of two situations or propositions.

The theory so far satisfies both of the requirements that we previously placed on situations. They are part of the causal order of the world, because they are taken to be pieces of reality, just as Barwise and Perry take real situations to be. They are in one-to-one correspondence with the true propositions, because they have been individuated in such a way that there is exactly one situation for every proposition that accords with reality. What may be in doubt, however, is that there will be enough propositions to do the work that notion normally does in semantics. Elsewhere (see Chapter 5), we show how the theory can be extended to handle first-order quantification, propositional connectives, and propositional attitude attributions, by admitting propositions and propositional functions among the entities to which properties and relations can be applied.

To summarize the extensions briefly: Propositional connectives become properties of propositions. Negation, for example, would be a unary property of propositions. A proposition has the negation property just in case it is false. For every false proposition, there is an actual situation of it being false, and for every proposition there is the additional proposition that it is false. Conjunction, disjunction, etc., become binary relations between propositions. First-order quantifiers become properties of functions from individuals to propositions.[4] For example, in standard logic "All men are mortal" is rendered as "Everything is such that, if it is a man, then it is mortal." In our framework this would be analyzed as the proposition: every individual is mapped into a true proposition by the function that maps an entity into the proposition that, if the entity is a man, then it is mortal.

Within this theory there is a natural semantics for first-order logic with formulas taken to denote propositions, with distinct formulas denoting distinct propositions unless they can be made identical by renaming of variables. We will therefore use the notation of standard

[4] A generalized quantifier treatment where quantifiers are considered to be binary relations on pairs of properties is probably preferable, but we present the simpler treatment in this chapter to be consistent with standard logic and with Davidson.

logic freely in the rest of this chapter, but with the semantics sketched here rather than the normal Tarskian semantics.

Situations and Events

The preceeding discussion makes an attempt to clarify the relation between situations and propositions, but what of events? Although we have claimed that situations are parts of the real world, they may seem rather abstract. Events, on the other hand, may seem much more real and familiar. For instance, if a bomb goes off, there seems little doubt that there really is such a thing as the explosion. We can see it and feel it, and it has undoubted causal effects. We will maintain, however, that situations and events are intimately related; that, in fact, robust large-scale events such as explosions consist of nothing more than the sum of (literally) uncountably many simple situations.

Suppose an object moves from point P_1 to point P_2 between T_1 and T_2. Consider the situation of the object being at P_1 at T_1, the situation of it being at P_2 at T_2, and all of the situations of it being at some intermediate point at the corresponding intermediate time. We claim that the event of the object moving from P_1 to P_2 between T_1 and T_2 consists of nothing more than the sum of all these situations. The argument is really quite simple: If all these situations exist—that is, if the object is at P_1 at T_1 and at P_2 at T_2 and at all the intermediate points at the corresponding intermediate times—then the movement in question exists. Nothing more needs to be added to these states of affairs for the moving event to exist; therefore it is gratuitous to assert that the moving event consists of anything beyond these situations.

The only qualification that needs to be mentioned is that the verb "consist" is used quite deliberately here, instead of the "be" of identity. That is because, according to common sense, one and the same event could have consisted of slightly different smaller events, and hence of a slightly different set of situations. World War II would not have been a different war merely if one fewer soldier had been killed. But this is no different than the observation that changing one screw on a complex machine does not make it a different machine. Thus we will say that situations are the stuff out of which events are made, just as material substances are the stuff out of which objects are made. The exact identity criteria for events in terms of situations are likely to be just as hard to define as for objects in terms of their material. But by the same token, there is no reason to conclude that there is something to an event over and above the situations it includes, any more than there is to conclude that there is something to an object over and above the material it is made of.

9.4 The Analysis

With this framework behind us, let us look again at "Jones buttered the toast." Perry begins his analysis by saying

> "Jones" refers to Jones, "the toast" refers to some piece of toast, and "buttered" refers to a relational activity, with the tense constraining the location (Perry 1983, p. 2).

This certainly seems unobjectionable. We have two objects and a binary relation, ignoring tense, as we do throughout this chapter. If the objects in question actually satisfy the relation, then there is a corresponding situation. But how is this situation related to the commonsense event of Jones buttering the toast? The buttering event is surely a complex motion, so by the argument of the last section it must consist of countless situations of the butter, the toast, the knife, Jones's arm, etc. being in certain positions at certain times. According to the identity criterion we have given for situations, those situations and the event which is constituted by their sum are distinct from the single situation of the buttering relation holding between Jones and the toast.

Clearly the buttering situation and the buttering event are closely related, but according to the principles we have adopted, they cannot be one and the same. Davidson's analysis of event sentences turns out to provide a very attractive way of expressing the relation between them. If we analyze an event sentence as asserting the existence of an event, as he suggests, then according to our semantic framework, the sentence asserts that a certain property of events is instantiated.[5] In the buttering toast example, the sentence says that the property of being a buttering of the toast by Jones is instantiated. The situation that the whole sentence describes, then, is the situation of the property of being a buttering of the toast by Jones being instantiated. Thus, on the one hand, we have a situation of a certain property of events being instantiated, and on the other hand we have the event that actually instantiates the property.

On first exposure, this may seem like an artificial distinction imposed to solve an artificial problem. In point of fact, however, this distinction is exactly what is needed to explain the two types of adverbial modification discussed in Section 9.2. Moreover, all the data presented there can then be quite straightforwardly accounted for within the framework we have developed.

[5] Strictly speaking, the theory says the sentence asserts there is an event mapped into a true proposition by a certain propositional function, but for simplicity we will paraphrase this in terms of the corresponding property of events.

Let us look again at perhaps the simplest pair of sentences illustrating these two types of modification:

(12) (a) John sang strangely.
(b) Strangely, John sang.

The manner use of the adverb in (12a) seems to fit quite comfortably within the Davidsonian pattern of treating adverbs as making additional predications of the event whose existence is asserted by the basic sentence. If John sang strangely, it seems most definitely to be the singing event itself that is strange. With (12b), though, the singing event itself may be quite ordinary as singing events go. It seems to be the fact that there is any singing by John at all that is strange. But this is precisely what we are saying if we analyze (12b) as predicating strangeness of the situation of the property of being-a-singing-by-John being instantiated.

We can represent this symbolically by making a minor extension to ordinary logic; (12a) can be represented in the way Davidson has already suggested.

(28) $\exists x(\mathsf{Sang}(\mathsf{John},\ x) \wedge \mathsf{Strange}(x))$

The extension is required to represent the fact use of the adverb in (12b). That sentence attributes strangeness to a situation, and since we have decided to let formulas denote propositions, we do not yet have any notation for situations. One remedy for this is to let situations be in the domain of individuals, as Davidson already assumes events to be, and to introduce a relation "Fact" that holds between a situation and the corresponding true proposition. The name "Fact" is chosen because this relation quite plausibly provides the semantics of the locution "the fact that *P*." Note that while "Fact" denotes a relation between a situation and a proposition in our semantics, it will be an operator whose first argument is a singular term and whose second argument is a formula, rather than an ordinary relation symbol. (12b) would then be represented by

(29) $\exists y(\mathsf{Fact}(y, \exists x(\mathsf{Sang}(\mathsf{John},\ x))) \wedge \mathsf{Strange}(y))$

This says literally that there exists a fact (or situation) of there being a singing-by-John event and that fact is strange, or more informally, the fact that John sang is strange.

If there is a distinct situation corresponding to every true proposition, it may be worrying to allow situations into the domain of individuals. There are various foundational approaches that could be used to justify this, but we will merely note that the logical principles needed for our use of situations are so weak that no inconsistency seems threatened. The only general principle that seems appropriate is the schema

(30) $\exists y(\mathsf{Fact}(y, P)) \equiv P$

This schema can easily be shown to be consistent by giving "Fact" a simple syntactic interpretation that makes the schema true.

Under this analysis of event sentences and adverbial modification, all the other data are easily explained. The factivity of the fact use of adverbs and their related adjectives arises because the adverbs and adjectives express properties of situations, which are real pieces of the world that do not exist unless the corresponding propositions are true.

Copular sentences do not exhibit the fact/manner distinction in their adverbial modifiers, because they do not involve event variables; only the overall situation is available for the adverb to be predicated of. This provides one answer to Perry's objection to the Davidsonian treatment of event sentences: "The idea that 'Sarah was walking' gets a cosmically different treatment than 'Sarah was agile' strikes me as not very plausible." (Perry 1983, p. 3) The first of these can take manner adverbials, and the second cannot, a fact that seems to require *some* difference in analysis to explain.

The extensionality with respect to noun phrases of sentences with manner adverbials follows directly from Davidson's original proposal. The noun phrases do not occur within the adverbial's ultimate scope, which is only the event variable. Changing the entire sentence, as in (24) through (27), changes the event, though, so we do not get that sort of extensionality.

The nonextensionality of sentences with fact adverbials follows from the fact that changing a description of a participant in an event changes the particular property of the event that goes into determining what situation is being discussed, even though the event itself does not change. If we compare (20) and (21),

(20) Rudely, John spoke to the Queen.

(21) Rudely, John spoke to the woman next to him.

we see that the two sentences describe a single event, John's speaking to the Queen, who is also the woman next to him. The sentences describe the event in two different ways, though, so they ascribe two different properties to it.[6] If we leave out the adverb, the unmodified sentences assert that these two properties of events are instantiated.

[6] To make sure these two properties do come out nonidentical in our semantics, we need to treat "the" as a quantifier. There are many independent reasons for doing this, however.

Since these properties are different, the situation of one of them being instantiated is a different situation from that of the other one being instantiated. Hence one of those situations might be rude (of John) without the other one being so.

9.5 Conclusions

Let us return to Davidson's and Perry's analyses of event sentences, to see how they fare in the light of the data and theory presented here. We have adopted Davidson's analysis of manner adverbials wholesale, so we are in complete agreement with him on that point. We sharply disagree with him, however, on the possibility of associating event-like entities, i.e., situations, with whole sentences, and we find them absolutely necessary to account for the fact use of adverbs, a case Davidson fails to consider. Perry, on the other hand, rightly takes Davidson to task for his faulty argument against associating situations with whole sentences, but then fails to look closely enough at the data to see that something like Davidson's analysis is still needed to account for the detailed facts about manner adverbials.

10

Unification-Based Semantic Interpretation

10.1 Introduction

Over the past several years, unification-based formalisms (Shieber 1986) have come to be widely used for specifying the syntax of natural languages, particularly among computational linguists. It is less widely realized by computational linguists that unification can also be a powerful tool for specifying the semantic interpretation of natural languages. This chapter shows how unification can be used to specify the semantic interpretation of natural-language expressions, including problematical constructions involving long-distance dependencies. The chapter also sketches a theoretical foundation for unification-based semantic interpretation, and it compares the unification-based approach with more conventional techniques based on the lambda calculus.

Many of the techniques described here are fairly well known among natural-language researchers working with logic grammars, but have not been extensively discussed in the literature, perhaps the only systematic presentation being that of Pereira and Shieber (1987). This chapter goes into many issues in greater detail than do Pereira and Shieber, however, and sketches what may be the first theoretical analysis of unification-based semantic interpretation.

The research reported in this chapter was begun at SRI International's Cambridge Computer Science Research Centre in Cambridge, England, supported by a grant from the Alvey Directorate of the U.K. Department of Trade and Industry and by the members of the NATTIE consortium (British Aerospace, British Telecom, Hewlett Packard, ICL, Olivetti, Philips, Shell Research, and SRI). The work was continued at the SRI Artificial Intelligence Center and the Center for the Study of Language and Information, supported in part by a gift from the Systems Development Foundation and in part by a contract with the Nippon Telegraph and Telephone Corporation.

We begin by reviewing the basic ideas behind unification-based grammar formalisms, which will also serve to introduce the style of notation to be used throughout the chapter. The notation is that used in the Core Language Engine (CLE) developed by SRI's Cambridge Computer Science Research Center in Cambridge, England, a system whose semantic-interpretation component makes use of many of the ideas presented here (Alshawi 1992).

Fundamentally, unification grammar is a generalization of context-free phrase structure grammar in which grammatical-category expressions are not simply atomic symbols, but have sets of features with constraints on their values. Such constraints are commonly specified using sets of equations. Our notation uses equations of a very simple format—just `feature=value`—and permits only one equation per feature per constituent, but we can indicate constraints that would be expressed in other formalisms using more complex equations by letting the value of a feature contain a variable that appears in more than one equation. The CLE is written in Prolog, to take advantage of the efficiency of Prolog unification in implementing category unification, so our grammar rules are written as Prolog assertions, and we follow Prolog conventions in that constants, such as category and feature names, start with lowercase letters, and variables start with uppercase letters. As an example, a simplified version of the rule for the basic subject-predicate sentence form might be written in our notation as

```
(1) syn(s_np_vp,
        [s:[type=tensed],
         np:[person=P,num=N],
         vp:[type=tensed,person=P,num=N]]).
```

The predicate `syn` indicates that this is a syntax rule, and the first argument `s_np_vp` is a rule identifier that lets us key the semantic-interpretation rules to the syntax rules. The second argument of `syn` is a list of category expressions that make up the content of the rule, the first specifying the category of the mother constituent and the rest specifying the categories of the daughter constituents. This rule, then, says that a tensed sentence (`s:[type=tensed]`) can consist of a noun phrase (`np`) followed by a verb phrase (`vp`), with the restrictions that the verb phrase must be tensed (`type=tensed`), and that the noun phrase and verb phrase must agree in person and number—that is, the `person` and `num` features of the noun phrase must have the same respective values as the `person` and `num` features of the verb phrase.

These constraints are checked in the process of parsing a sentence by unifying the values of features specified in the rule with the values of features in the constituents found in the input. Suppose, for instance, that we are parsing the sentence *Mary runs* using a left-corner parser. If *Mary* is parsed as a constituent of category

```
np:[person=3rd,num=sing],
```

then unifying this category expression with

```
np:[person=P,num=N]
```

in applying the sentence rule above will force the variables `P` and `N` to take on the values `3rd` and `sing`, respectively. Thus when we try to parse the verb phrase, we know that it must be of the category

```
vp:[type=tensed,person=3rd,num=sing].
```

Our notation for semantic-interpretation rules is a slight generalization of the notation for syntax rules. The only change is that in each position where a syntax rule would have a category expression, a semantic rule has a pair consisting of a "logical-form" expression and a category expression, where the logical-form expression specifies the semantic interpretation of the corresponding constituent. A semantic-interpretation rule corresponding to syntax rule (1) might look like the following:

```
(2) sem(s_np_vp,
        [(apply(Vp,Np),s:[]),
         (Np,np:[]),
         (Vp,vp:[])]).
```

The predicate `sem` means that this is a semantic-interpretation rule, and the rule identifier `s_np_vp` indicates that this rule applies to structures built by the syntax rule with the same identifier. The list of pairs of logical-form expressions and category expressions specifies the logical form of the mother constituent in terms of the logical forms and feature values of the daughter constituents. In this case the rule says that the logical form of a sentence generated by the `s_np_vp` rule is an applicative expression with the logical form of the verb phrase as the functor and the logical form of the noun phrase as the argument. (The dummy functor `apply` is introduced because Prolog syntax does not allow variables in functor position.) Note that there are no feature restrictions on any of the category expressions occurring in the rule. They are unnecessary in this case because the semantic rule applies only to structures built by the `s_np_vp` syntax rule, and thus inherits all the restrictions applied by that rule.

10.2 Functional Application vs. Unification

Example (2) is typical of the kind of semantic rules used in the standard approach to semantic interpretation in the tradition established by Richard Montague (1974b, Dowty, Wall, and Peters 1981). In this approach, the interpretation of a complex constituent is the result of the functional application of the interpretation of one of the daughter constituents to the interpretation of the others.

A problem with this approach is that if, in a rule like (2), the verb phrase itself is semantically complex, as it usually is, a lambda expression has to be used to express the verb-phrase interpretation, and then a lambda reduction must be applied to express the sentence interpretation in its simplest form (Dowty, Wall, and Peters 1981, p. 98–111). To use (2) to specify the interpretation of the sentence *John likes Mary*, the logical form for *John* could simply be `john`, but the logical form for *likes Mary* would have to be something like `X\like(X,mary)`. (The notation `Var\Body` for lambda expressions is borrowed from Lambda Prolog (Miller and Nadathur 1986).) The logical form for the whole sentence would then be `apply(X\like(X,mary),john)`, which must be reduced to yield the simplified logical form `like(john,mary)`.

Moreover, lambda expressions and the ensuing reductions would have to be introduced at many intermediate stages if we wanted to produce simplified logical forms for the interpretations of complex constituents such as verb phrases. If we want to accommodate modal auxiliaries, as in *John might like Mary*, we have to make sure that the verb phrase *might like Mary* receives the same type of interpretation as *like(s) Mary* in order to combine properly with the interpretation of the subject. If we try to maintain functional application as the only method of semantic composition, then it seems that the simplest logical form we can come up with for *might like Mary* is produced by the following rule:

```
(3) sem(vp_aux_vp,
        [(X\apply(Aux,apply(Vp,X)),vp:[]),
         (Aux,aux:[]),
         (Vp,vp:[])]).
```

Applying this rule to the simplest plausible logical forms for *might* and *like Mary* would produce the following logical form for *might like Mary*:

```
X\apply(might,(apply(Y\like(Y,mary),X)))
```

which must be reduced to obtain the simpler expression

```
X\might(like(X,mary))
```

When this expression is used in the sentence-level rule, another reduction is required to eliminate the remaining lambda expression. The part of the reduction step that gets rid of the `apply` functors is to some extent an artifact of the way we have chosen to encode these expressions as Prolog terms, but the lambda reductions are not. They are inherent in the approach, and normally each rule will introduce at least one lambda expression that needs to be reduced away.

It is, of course, possible to add a lambda-reduction step to the interpreter for the semantic rules, but it is both simpler and more efficient to use the feature system and unification to do explicitly what lambda expressions and lambda reduction do implicitly—assign a value to a variable embedded in a logical-form expression. According to this approach, instead of the logical form for a verb phrase being a logical predicate, it is the same as the logical form of an entire sentence, but with a variable as the subject argument of the verb and a feature on the verb phrase having that same variable as its value. The sentence interpretation rule can thus be expressed as

```
(4) sem(s_np_vp,
        [(Vp,s:[]),
         (Np,np:[]),
         (Vp,vp:[subjval=Np])]),
```

which says that the logical form of the sentence is just the logical form of the verb phrase with the subject argument of the verb phrase unified with the logical form of the subject noun phrase. If the verb phrase *likes Mary* is assigned the logical-form/category-expression pair

```
(like(X,mary),vp:[subjval=X]),
```

then the application of this rule will unify the logical form of the subject noun phrase, say `john`, directly with the variable `X` in `like(X,mary)` to immediately produce a sentence constituent with the logical form `like(john,mary)`.

Modal auxiliaries can be handled equally easily by a rule such as

```
(5) sem(vp_aux_vp,
        [(Aux,vp:[subjval=S]),
         (Aux,aux:[argval=Vp]),
         (Vp,vp:[subjval=S])]).
```

If *might* is assigned the logical-form/category-expression pair

```
(might(A),aux:[argval=A]),
```

then applying this rule to interpret the verb phrase *might like Mary* will unify `A` in `might(A)` with `like(X,mary)` to produce a constituent with the logical-form/category-expression pair

```
(might(like,X,mary),vp:[subjval=X]),
```

which functions in the sentence-interpretation rule in exactly the same way as the logical-form/category-expression pair for *like Mary*.

10.3 Are Lambda Expressions Ever Necessary?

The approach presented above for eliminating the explicit use of lambda expressions and lambda reductions is quite general, but it does not replace all possible uses of lambda expressions in semantic interpretation. Consider the sentence *John and Bill like Mary*. The simplest logical form for the distributive reading of this sentence would be

```
and(like(john,mary),like(bill,mary)).
```

If the verb phrase is assigned the logical-form/category-expression pair

```
(like(X,mary),vp:[subjval=X]),
```

as we have suggested, then we have a problem: Only one of `john` or `bill` can be directly unified with `X`, but to produce the desired logical form, we seem to need two instances of `like(X,mary)`, with two different instantiations of `X`.

Another problem arises when a constituent that normally functions as a predicate is used as an argument instead. Common nouns, for example, are normally used to make direct predications, so a noun like *senator* might be assigned the logical-form/category-expression pair

```
(senator(X),nbar:[argval=X])
```

according to the pattern we have been following. (Note that we do not have "noun" as a syntactic category; rather, a common noun is simply treated as a lexical "n-bar.") It is widely recognized, however, that there are "intensional" adjectives and adjective phrases, such as *former*, that need to be treated as higher-level predicates or operators on predicates, so that in an expression like *former senator*, the noun *senator* is not involved in directly making a predication, but instead functions as an argument to *former*. We can see that this must be the case, from the observation that a former senator is no longer a senator. The logical form we have assigned to *senator*, however, is not literally that of a predicate, however, but rather of a complete formula with a free variable. We therefore need some means to transform this formula with its free variable into an explicit predicate to be an argument of *former*. The introduction of lambda expressions provides the solution to this problem, because the transformation we require is exactly what is accomplished by lambda abstraction. The following rule shows how this can be carried out in practice:

```
(6) sem(nbar_adj_nbar,
        [(Adjp,nbar:[argval=A]),
         (Adjp,adjp:[type=intensional,argval1=X\Nbar,
                     argval2=A]),
         (Nbar,nbar:[argval=X])]).
```

This rule requires that the logical-form/category-expression pair assigned to an intensional adjective phrase be something like

```
(former(P,Y),adjp:[type=intensional,argval1=P,
                   argval2=Y]),
```

where `former(P,Y)` means that `Y` is a former `P`. The daughter `nbar` is required to be as previously supposed. The rule creates a lambda expression, by unifying the bound variable with the argument of the daughter `nbar` and making the logical form of the daughter `nbar` the body of the lambda expression, and unifies the lambda expression with the first argument of the `adjp`. The second argument of the `adjp` becomes the argument of the mother `nbar`. Applying this rule to *former senator* will thus produce a constituent with the logical-form/category-expression pair

```
(former(X\senator(X),Y),nbar:[argval=Y]).
```

This solution to the second problem also solves the first problem. Even in the standard lambda-calculus-based approach, the only way in which multiple instances of a predicate expression applied to different arguments can arise from a single source is for the predicate expression to appear as an argument to some other expression that contains multiple instances of that argument. Since our approach requires turning a predicate into an explicit lambda expression if it is used as an argument, by the time we need multiple instances of the predicate, it is already in the form of a lambda expression. We can show how this works by encoding a Montagovian (Dowty, Wall, Peters 1981) treatment of conjoined subject noun phrases within our approach. The major feature of this treatment is that noun phrases act as higher-order predicates of verb phrases, rather than the other way around as in the simpler rules presented in Sections 10.1 and 10.2. In the Montagovian treatment, a proper noun such as *John* is given an interpretation equivalent to `P\P(john)`, so that when we apply it to a predicate like `run` in interpreting *John runs* we get something like `apply(P\P(john),run)` which reduces to `run(john)`. With this in mind, consider the following two rules for the interpretation of sentences with conjoined subjects:

```
(7) sem(np_np_conj_np
        [(Conj,np:[argval=P]),
         (Np1,np:[argval=P]),
         (Conj,conj:[argval1=Np1,argval2=Np2]),
         (Np2,np:[argval=P])]).
(8) sem(s_np_vp,
        [(Np,s:[]),
         (Np,np:[argval=X\Vp]),
         (Vp,vp:[subjval=X])]).
```

The first of these rules gives a Montagovian treatment of conjoined noun phrases, and the second gives a Montagovian treatment of simple declarative sentences. Both of these rules assume that a proper noun such as *John* would have a logical-form/category-expression pair like

```
(apply(P,john),np:[argval=P]).
```

In (7) it is assumed that the conjunction *and* would have a logical-form/category-expression pair like

```
(and(P1,P2),conj:[argval1=P1,argval2=P2]).
```

In (7) the logical forms of the two conjoined daughter `nps` are unified with the two arguments of the conjunction, and the arguments of the daughter `nps` are unified with each other and with the single argument of the mother `np`. Thus applying (7) to interpret *John and Bill* yields a constituent with the logical-form/category-expression pair

```
(and(apply(P,john),apply(P,bill)),np:[argval=P]).
```

In (8) an explicit lambda expression is constructed out of the logical form of the `vp` daughter in the same way a lambda expression was constructed in (6), and this lambda expression is unified with the argument of the subject `np`. For the sentence *John and Bill like Mary*, this would produce the logical form

```
and(apply(X\like(X,mary),john),
    apply(X\like(X,mary),bill)),
```

which can be reduced to `and(like(john,mary),like(bill,mary))`.

10.4 Theoretical Foundations of Unification-Based Semantics

The examples presented above ought to be convincing that a unification-based formalism can be a powerful tool for specifying the interpretation of natural-language expressions. What may not be clear is whether there is any reasonable theoretical foundation for this approach, or whether it is just so much unprincipled "feature hacking."

The informal explanations we have provided of how particular rules work, stated in terms of unifying the logical form for constituent X with the appropriate variable in the logical form for constituent Y, may suggest that the latter is the case. If no constraints are placed on how such a formalism is used, it is certainly possible to apply it in ways that have no basis in any well-founded semantic theory. Nevertheless, it is possible to place restrictions on the formalism to ensure that the rules we write have a sound theoretical basis, while still permitting the sorts of rules that seem to be needed to specify the semantic interpretation of natural languages.

The main question that arises in this regard is whether the semantic rules specify the interpretation of a natural-language expression in a compositional fashion. That is, does every rule assign to a mother constituent a well-defined interpretation that depends solely on the interpretations of the daughter constituents? If the interpretation of a constituent is taken to be just the interpretation of its logical-form expression, the answer is clearly "no." In our formalism the logical-form expression assigned to a mother constituent depends on both the logical-form expressions and the category expressions assigned to its daughters. As long as both category expressions and logical-form expressions have a theoretically sound basis, however, there is no reason that both should not be taken into account in a semantic theory; so, we will define the interpretation of a constituent based on both its category and its logical form. Taking the notion of interpretation in this way, we will explain how our approach can be made to preserve compositionality. First, we will show how to give a well-defined interpretation to every constituent; then, we will sketch the sort of restrictions on the formalism one needs to guarantee that any interpretation-preserving substitution for a daughter constituent also preserves the interpretation of the mother constituent.

The main problem in giving a well-defined interpretation to every constituent is how to interpret a constituent whose logical-form expression contains free variables that also appear in feature values in the constituent's category expression. Recall the rule we gave for combining auxiliaries with verb phrases:

```
(5) sem(vp_aux_vp,
        [(Aux,vp:[subjval=S]),
         (Aux,aux:[argval=Vp]),
         (Vp,vp:[subjval=S])]).
```

This rule accepts daughter constituents having logical-form/category-expression pairs such as

```
(might(A),aux:[argval=A])
```

and

```
(like(X,mary),vp:[subjval=X])
```

to produce a mother constituent having the logical-form/category-expression pair

```
(might(like,X,mary),vp:[subjval=X].
```

Each of these pairs has a logical-form expression containing a free variable that also occurs as a feature value in its category expression. The simplest way to deal with logical-form/category-expression pairs such as these is to regard them in the way that syntactic-category expressions in unification grammar can be regarded—as abbreviations for the set of all their well-formed fully instantiated substitution instances.

To establish some terminology, we will say that a logical-form/category-expression pair containing no free-variable occurrences has a "basic interpretation," which is simply the ordered pair consisting of the interpretation of the logical-form expression and the interpretation of the category expression. Since there are no free variables involved, basic interpretations should be unproblematic. The logical-form expression will simply be a closed well-formed expression of some ordinary logical language, and its interpretation will be whatever the usual interpretation of that expression is in the relevant logic. The category expression can be taken to denote a fully instantiated grammatical category of the sort typically found in unification grammars. The only unusual property of this category is that some of its features may have logical-form interpretations as values, but, as these will always be interpretations of expressions containing no free-variable occurrences, they will always be well defined.

Next, we define the interpretation of an arbitrary logical-form/category-expression pair to be the set of basic interpretations of all its well-formed substitution instances that contain no free-variable occurrences. For example, the interpretation of a constituent with the logical-form/category-expression pair

```
(might(like,X,mary),vp:[subjval=X])
```

would consist of a set containing the basic interpretations of such pairs as

```
(might(like,john,mary),vp:[subjval=john]),
(might(like,bill,mary),vp:[subjval=bill]),
```

and so forth.

This provides well-defined interpretation for every constituent, so we can now consider what restrictions we can place on the formalism to guarantee that any interpretation-preserving substitution for a daughter constituent also preserves the interpretation of its mother constituent. The first restriction we need rules out constituents that would have degenerate interpretations: No semantic rule or semantic lexical specification may contain both free and bound occurrences of the same variable in a logical-form/category-expression pair.

To see why this restriction is needed, consider the logical-form/category-expression pair

```
(every(X,man(X),die(X)),np:[boundvar=X,bodyval=die(X)]),
```

which might be the substitution instance of a daughter constituent that would be selected in a rule that combines noun phrases with verb phrases. The problem with such a pair is that it does not have any well-formed substitution instances that contain no free-variable occurrences. The variable `X` must be left uninstantiated in order for the logical-form expression `every(X,man(X),die(X))` to be well formed, but this requires a free occurrence of `X` in `np:[boundvar=X,bodyval=die(X)]`. Thus this pair will be assigned the empty set as its interpretation. Since any logical-form/category-expression pair that contains both free and bound occurrences of the same variable will receive this degenerate interpretation, any other such pair could be substituted for this one without altering the interpretations of the daughter constituent substitution instances that determine the interpretation of the mother constituent. It is clear that this would normally lead to gross violations of compositionality, since the daughter substitution instances selected for the noun phrases *every man*, *no woman*, and *some dog* would all receive the same degenerate interpretation under this scheme.

This restriction may appear to be so constraining as to rule out certain potentially useful ways of writing semantic rules, but in fact it is generally possible to rewrite such rules in ways that do not violate the restiction. For example, in place of the sort of logical-form/category-expression pair we have just ruled out, we can fairly easily rewrite the relevant rules to select daughter substitution instances such as

```
(every(X,man(X),die(X)),np:[bodypred=X\die(X)]),
```

which does not violate the constraint and has a completely straightforward interpretation.

Having ruled out constituents with degenerate interpretations, the principal remaining problem is how to exclude rules that depend on properties of logical-form expressions over and above their interpretations. For example, suppose that the order of conjuncts does not affect the interpretation of a logical conjunction, according to the interpretation of the logical-form language. That is, `and(p,q)` would have the same interpretation as `and(q,p)`. The potential problem that this raises is that we might write a semantic rule that contains both a logical-form expression like `and(P,Q)` in the specification of a daughter constituent and the variable `P` in the logical form of the mother constituent. This would be a violation of compositionality, because the interpretation of the mother would depend on the interpretation of the left conjunct of a conjunction, even though, according to the semantics of the logical-form language, it makes no sense to distinguish the left and right conjuncts. If order of conjunction does not affect meaning, we ought to be able to substitute a daughter with the logical form `and(q,p)` for one with the logical form `and(p,q)` without affecting the interpretation assigned to the mother, but clearly, in this case, the interpretation of the mother would be affected.

It is not clear that there is any uniquely optimal set of restrictions that guarantees that such violations of compositionality cannot occur. Indeed, since unification formalisms in general have Turing machine power, it is quite likely that there is no computable characterization of all and only the sets of semantic rules that are compositional. Nevertheless, one can describe sets of restrictions that do guarantee compositionality, and which seem to provide enough power to express the sorts of semantic rules we need to use to specify the semantics of natural languages. One fairly natural way of restricting the formalism to guarantee compositionality is to set things up so that unifications involving logical-form expressions are generally made against variables, so that it is possible neither to extract subparts of logical-form expressions nor to filter on the syntactic form of logical-form expressions. The only exception to this restriction that seems to be required in practice is to allow for rules that assemble and disassemble lambda expressions with respect to their bodies and bound variables. So long as no extraction from inside the body of a lambda expression is allowed, however, compositionality is preserved.

It is possible to define a set of restrictions on the form of semantic rules that guarantee that no rule extracts subparts (other than the body or bound variable of a lambda expression) of a logical-form expression or filters on the syntactic form of a logical-form expression. The statement of these restrictions is straightforward, but rather long and tedious, so we omit the details here. We will simply note that none of the sample rules presented in this chapter involve any such extraction or filtering.

10.5 Semantics of Long-Distance Dependencies

The main difficulty that arises in formulating semantic-interpretation rules is that constituents frequently appear syntactically in places that do not directly reflect their semantic role. Semantically, the subject of a sentence is one of the arguments of the verb, so it would be much easier to produce logical forms for sentences if the subject were part of the verb phrase. The use of features such as **subjval**, in effect, provides a mechanism for taking the interpretation of the subject from the place where it occurs and inserting it into the verb phrase interpretation where it "logically" belongs.

The way features can be manipulated to accomplish this is particularly striking in the case of the long-distance dependencies, such as those in WH-questions. For the sentence *Which girl might John like?*, the simplest plausible logical form would be something like

```
which(X,girl(X),might(like(john,X))),
```

where the question-forming operator **which** is treated as a generalized quantifier whose "arguments" consist of a bound variable, a restriction, and a body.

The problem is how to get the variable **X** to link the part of the logical form that comes from the fronted interrogative noun phrase with the argument of **like** that corresponds to the noun phrase gap at the end of the verb phrase. To solve this problem, we can use a technique called "gap-threading." This technique was introduced in unification grammar to describe the syntax of constructions with long-distance dependencies (Karttunnen 1986, Pereira and Shieber 1987, p. 125–129), but it works equally well for specifying their semantics. The basic idea is to use a pair of features, **gapvalsin** and **gapvalsout**, to encode a list of semantic "gap fillers" to be used as the semantic interpretations of syntactic gaps, and to thread that list along to the points where the gaps occur. These gap fillers are often just the bound variables introduced by the constructions that permit gaps to occur.

The following semantic rules illustrate how this mechanism works:

```
(9) sem(whq_ynq_np_gap,
        [(Np,s:[gapvalsin=[],gapvalsout=[]]),
         (Np,np:[type=interrog,bodypred=A\Ynq]),
         (Ynq,s:[gapvalsin=[A],gapvalsout=[]])]).
```

This is the semantic-interpretation rule for a WH-question with a long-distance dependency. The syntactic form of such a sentence is an interrogative noun phrase followed by a yes/no question with a noun phrase gap. This rule expects the interrogative noun phrase *which girl* to have a logical-form/category-expression pair such as

```
(which(X,girl(X),Bodyval),
 np:[type=interrog,bodypred=X\Bodyval]).
```

The feature **bodypred** holds a lambda expression whose body and bound variable are unified respectively with the body and the bound variable of the **which** expression. In (9) the body of this lambda expression is unified with the logical form of the embedded yes/no question, and the **gapvalsin** feature is set to be a list containing the bound variable of the lambda expression. This list is actually used as a stack, to accomodate multiply nested filler-gap dependencies. Since this form of question cannot be embedded in other constructions, however, we know that in this case there will be no other gap-fillers already on the list.

This is the rule that provides the logical form for empty noun phrases:

```
(10)  sem(empty_np,
          [(Val,np:[gapvalsin=[Val|ValRest],
                    gapvalsout=ValRest])]).
```

Notice that it has a mother category, but no daughter categories. The rule simply says that the logical form of an empty **np** is the first element on its list of semantic gap-fillers, and that this element is "popped" from the gap-filler list. That is, the **gapvalsout** feature takes as its value the tail of the value of the **gapvalsin** feature.

We now show two rules that illustrate how a list of gap-fillers is passed along to the points where the gaps they fill occur.

```
(11)  sem(vp_aux_vp,
          [(Aux,vp:[subjval=S,gapvalsin=In,
                    gapvalsout=Out]),
           (Aux,aux:[argval=Vp]),
           (Vp,vp:[subjval=S,gapvalsin=In,
                   gapvalsout=Out])]).
```

This semantic rule for verb phrases formed by an auxilliary followed by a verb phrase illustrates the typical use of the gap features to "thread" the list of gap fillers through the syntactic structure of the sentence to the points where they are needed. An auxillary verb cannot be or contain a WH-type gap, so there are no gap features on the category **aux**. Thus the gap features on the mother **vp** are simply unified with the corresponding features on the daughter **vp**.

A more complex case is illustrated by the following rule:

(12)
```
sem(vp_vp_pp,
      [(Pp,vp:[subjval=S,gapvalsin=In,
               gapvalsout=Out]),
       (Vp,vp:[subjval=S,gapvalsin=In,
               gapvalsout=Across]),
       (Pp,pp:[argval=Vp,gapvalsin=Across,
               gapvalsout=Out])]).
```

This is a semantic rule for verb phrases that consist of a verb phrase and a prepositional phrase. Since WH-gaps can occur in either verb phrases or prepositional phrases, the rule threads the list carried by the **gapvalsin** feature of the mother **vp** first through the daughter **vp** and then through the daughter **pp**. This is done by unifying the mother **vp**'s **gapvalsin** feature with the daughter **vp**'s **gapvalsin** feature, the daughter **vp**'s **gapvalsout** feature with the daughter **pp**'s **gapvalsin** feature, and finally the daughter **pp**'s **gapvalsout** feature with the mother **vp**'s **gapvalsout** feature. Since a gap-filler is removed from the list once it has been "consumed" by a gap, this way of threading ensures that fillers and gaps will be matched in a last-in-first-out fashion, which seems to be the general pattern for English sentences with multiple filler-gap dependencies. (This does not handle "parasitic gap" constructions, but these are very rare and at present there seems to be no really convincing linguistic account of when such constructions can be used.)

Taken altogether, these rules push the quantified variable of the interrogative noun phrase onto the list of gap values encoded in the feature **gapvalsin** on the embedded yes/no-question. The list of gap values gets passed along by the gap-threading mechanism, until the empty-noun-phrase rule pops the variable off the gap values list and uses it as the logical form of the noun phrase gap. Then the entire logical form for the embedded yes/no-question is unified with the body of the logical form for the interrogative noun phrase, producing the desired logical form for the whole sentence.

This treatment of the semantics of long-distance dependencies provides us with an answer to the question of the relative expressive power of our approach compared with the conventional lambda-calculus-based approach. We know that the unification-based approach is at least as powerful as the conventional approach, because the the conventional approach can be embedded directly in it, as illustrated by the examples in Section 10.3. What about the other way around? Many unification-based rules have direct lambda-calculus-based counterparts; for example (2) is a counterpart of (4), and (3) is the counterpart of (5). Once we introduce gap-threading, however, the correspondence breaks down. In the conventional approach, each rule applies only to constituents whose semantic interpretation is of some particular single semantic type, say, functions from individuals to propositions. If every free variable in our approach is treated as a lambda variable in the conventional approach, then no one rule can cover two expressions whose interpretation essentially involves different numbers of variables, since these would be of different semantic types. Hence, rules like (11) and (12), which cover constituents containing any number of gaps, would have to be replaced in the conventional approach by a separate rule for each possible number of gaps. Thus, our formalism enables us to write more general rules than is possible taking the conventional approach.

10.6 Conclusions

In this chapter we have tried to show that a unification-based approach can provide powerful tools for specifying the semantic interpretation of natural-language expressions, while being just as well founded theoretically as the conventional lambda-calculus-based approach. Although the unification-based approach does not provide a substitute for all uses of lambda expressions in semantic interpretation, we have shown that lambda expressions can be introduced very easily where they are needed. Finally, the unification-based approach provides for a simpler statement of many semantic-interpretation rules, it eliminates many of the lambda reductions needed to express semantic interpretations in their simplest form, and in some cases it allows more general rules than can be stated taking the conventional approach.

References

Almog J., J. Perry and H. Wettstein. 1989. *Themes from Kaplan*, Oxford University Press, London, England.

Alshawi, H., ed. 1992. *The Core Language Engine*. Cambridge: MIT Press/Bradford Books.

Barwise, J. and J. Etchemendy. 1987. *The Liar*. New York: Oxford University Press.

Barwise, J. and J. Perry. 1983. *Situations and Attitudes*. Cambridge: MIT Press/Bradford Books.

Barwise, J. and J. Perry. 1985. Shifting Situations and Shaken Attitudes. *Linguistics and Philosophy* 8(1): 105–161.

Brooks, R. A. 1991a. Intelligence Without Representation. *Artificial Intelligence*. 47(1–3): 139–159.

Brooks, R. A. 1991b. Intelligence Without Reasoning. In *Proceedings of the 12th International Joint Conference on Artificial Intelligence*, 569–595. Sydney, Australia.

Chomsky, N. 1975. *Reflections on Language*. New York: Pantheon Books.

Colmerauer, A. 1978. Metamorphosis Grammars. In *Natural-language Communication with Computers*, ed. L. Bolc. Berlin: Springer-Verlag.

Cresswell, M. J. 1982. The Autonomy of Semantics. In *Processes, Beliefs, and Questions*, ed. S. Peters and E. Saarinen, 69–96. Dordrecht: D. Reidel Publishing Company.

Davidson, D. 1967a. Truth and Meaning. *Synthese* 17: 304–323.

Davidson, D. 1967b. The Logical Form of Action Sentences. In *The Logic of Decision and Action*, ed. N. Rescher, 81–95. Pittsburgh: University of Pittsburgh Press.

Dennett, D. C. 1978. *Brainstorms* Montgomery, Vermont: Bradford Books, Publishers.

Dowty, D. R., R. Wall, and S. Peters. 1981. *Introduction to Montague Semantics*. Dordrecht, Holland: D. Reidel.

Fodor, J. A. 1975. *The Language of Thought*. New York: Thomas Y. Crowell Company.

Frege, G. 1949. On Sense and Nominatum. In *Readings in Philosophical Analysis*, ed. H. Feigl and W. Sellars, 85–102. New York: Appleton-Century-Crofts, Inc.

Gelfond, M. 1987. On Stratified Autoepistemic Theories. In *Proceedings AAAI-87*, 207-211. Seattle, Washington.

Gelfond, M. and V. Lifschitz. 1988. The Stable Model Semantics for Logic Programming. In *Proceedings Fifth Logic Programming Symposium*, ed. R. Kowalski and K. Bowen, 1070–1080. Cambridge, Massachusetts: The MIT Press.

Halpern, J. Y. and Y. Moses. 1984. Towards a Theory of Knowledge and Ignorance: Preliminary Report. In *Proceedings Nonmonotonic Reasoning Workshop*, 125-143. Mohonk Mountain House, New Paltz, New York.

Hayes, P. J. 1973. Computation and Deduction. In *Proc. 2nd Symposium on Mathematical Foundations of Computer Science*, Czechoslovak Academy of Sciences, 105–116.

Hayes, P. J. 1977. In Defence of Logic. In *Proc. Fifth International Joint Conference on Artificial Intelligence*, 559–565. Cambridge, Massachusetts.

Hintikka, J. 1962. *Knowledge and Belief: An Introduction to the Logic of the Two Notions*. Ithaca: Cornell University Press.

Hintikka, J. 1971. Semantics for Propositional Attitudes. In *Reference and Modality*, ed. L. Linsky, 145–167. London: Oxford University Press.

Hughes, G. E. and M. J. Cresswell. 1968. *An Introduction to Modal Logic*. London: Methuen and Company, Ltd.

Kaplan, D. 1969. Quantifying In. In *Words and Objections: Essays on the Work of W. V. Quine*, ed. D. Davidson and J. Hintikka, 178–214. Dordrecht: D. Reidel Publishing Company.

Kaplan, K. 1977. Demonstratives: An Essay on the Semantics, Logic, Metaphysics, and Epistemology of Demonstratives and Other Indexicals. Unpublished manuscript, Department of Philosophy, University of California at Los Angeles.

Karttunnen, L. 1986. D-PATR: A Development Environment for Unification-Based Grammars. In *Proceedings of the 11th International Conference on Computational Linguistics*, Bonn, West Germany, 74–80.

Kintsch, W. 1974. *The Representation of Meaning in Memory* Hillsdale, New Jersey: Lawrence Erlbaum Associates, Inc.

Konolige, K. 1985. Belief and Incompleteness. In *Formal Theories of the Commonsense World*, ed. J. R. Hobbs and R. C. Moore, 359–403.

Konolige, K. 1988. On the Relation Between Default Logic and Autoepistemic Theories. *Artificial Intelligence* 35, No. 3, 343–382.

Konolige, K. 1991. Quantification in Autoepistemic Logic. *Fundamenta Informaticae*, 15, Nos. 3–4, 275–300.

Kowalski, R. 1974. Predicate Logic as a Programming Language. In *Information Processing 74*, 569–574. Amsterdam: North-Holland Publishing Company.

Kowalski, R. 1979. *Logic for Problem Solving*. New York: Elsevier North Holland, Inc.

Kripke, S. A. 1963. Semantical Analysis of Modal Logic. *Zeitschrift fuer Mathematische Logik und Grundlagen der Mathematik* 9: 67–96.

Kripke, S. A. 1971. Semantical Considerations on Modal Logic. In *Reference and Modality*, ed. L. Linsky, 63–72. London: Oxford University Press.

Kripke, S. A. 1972. Naming and Necessity. In *Semantics of Natural Language*, ed. D. Davidson and G. Harmon, 253–355. Dordrecht: D. Reidel Publishing Co.

Kripke, S. A. 1979. A Puzzle about Belief. In *Meaning and Use*, ed. A. Margalit, 239–283. Dordrecht: D. Reidel Publishing Company.

Levesque, H. J. 1981. The Interaction with Incomplete Knowledge Bases: A Formal Treatment. In *Proceedings of the Seventh International Joint Conference on Artificial Intelligence*. University of British Columbia, Vancouver, B.C., Canada, 240–245.

Levesque, H. 1990. All I Know: A Study in Autoepistemic Logic. *Artificial Intelligence* 42, Nos. 2–3, 263–309.

Lewis, D. 1972. General Semantics. In *Semantics of Natural Language*, ed. D. Davidson and G. Harmon, 169–218. Dordrecht: D. Reidel Publishing Co.

Lucas, J. R. 1961. Minds, Machines and Goedel. In *Minds and Machines*, ed. A. R. Anderson, 43–59. Englewood Cliffs, New Jersey: Prentice-Hall, Inc.

Marek, W. and M. Truszczynski. 1989. Relating Autoepistemic and Default Logic. In *Proceedings First International Conference on Principles of Knowledge Representation and Reasoning*, Toronto, Canada, 276–288.

McCarthy, J. 1962. Towards a Mathematical Science of Computation. In *Information Processing, Proceedings of IFIP Congress 62*, ed. C. Popplewell, 21–28. Amsterdam: North-Holland Publishing Company.

McCarthy, J. 1968. Programs with Common Sense. In *Semantic Information Processing*, ed. M. Minsky, 403–418. Cambridge, Massachusetts: The MIT Press.

McCarthy, J. and J. Hayes. 1969. Some Philosophical Problems from the Standpoint of Artificial Intelligence. In *Machine Intelligence 4*, ed. B. Meltzer and D. Michie, 463–502. Edinburgh: Edinburgh University Press.

McCarthy, J., et al. 1962. *LISP 1.5 Programmer's Manual* Cambridge, Massachusetts: The MIT Press.

McClelland, J. L., D. E. Rumelhart, and the PDP Research Group. 1986. *Parallel Distributed Processing: Explorations in the Microstructure of Cognition, Volume 2: Psychological and Biological Models*. Cambridge, Massachusetts: The MIT Press/Bradford Books.

McDermott, D. and J. Doyle. 1980. Non-Monotonic Logic I. *Artificial Intelligence* 13(1, 2): 41–72.

McDermott, D. 1982. Nonmonotonic Logic II: Nonmonotonic Modal Theories. *Journal of the Association for Computing Machinery* 29(1): 33–57.

Miller, D. A. , and G. Nadathur. 1986. Higher-Order Logic Programming. In *Third International Conference on Logic Programming*, ed. E. Shapiro, 448–462. Berlin: Springer-Verlag.

Minsky, M. 1974. A Framework for Representing Knowledge, MIT Artificial Intelligence Laboratory, AIM-306, Massachusetts Institute of Technology, Cambridge, Massachusetts.

Montague, R. 1974a. Pragmatics and Intensional Logic. In *Formal Philosophy, Selected Papers of Richard Montague*, ed. R. H. Thomason, 119–147. New Haven and London: Yale University Press.

Montague, R. 1974b. *Formal Philosophy: Selected Papers of Richard Montague*, ed. R. H. Thomason. New Haven: Yale University Press.

Montague, R. 1974c. English as a Formal Language. In *Formal Philosophy, Selected Papers of Richard Montague*, ed. R. H. Thomason, 188–221. New Haven and London: Yale University Press.

Moore, R. C. 1980a. Reasoning About Knowledge and Action, Artificial Intelligence Center Technical Note 191, SRI International, Menlo Park, California.

Moore, R. C. 1980b. *Reasoning from Incomplete Knowledge in a Procedural Deduction System*. New York: Garland Publishing, Inc.

Moravcsik, J. 1973. Comments on Partee's Paper. In *Approaches to Natural Language*, ed. J. K. K. Hintikka et al., 349–369. Dordrecht: D. Reidel Publishing Co.

Newell, A. 1980. The Knowledge Level, Presidential Address, American Association for Artificial Intelligence, AAAI80, Stanford University, Stanford, California. In *AI Magazine* 2(2)1–20.

Nilsson, N. J. 1980. *Principles of Artificial Intelligence*. Palo Alto: Tioga Publishing Company.

Partee, B. H. 1973. The Semantics of Belief-Sentences. In *Approaches to Natural Language*, ed. J. K. K. Hintikka et al., 309–336. Dordrecht: D. Reidel Publishing Co.

Partee, B. H. 1979. Semantics—Mathematics or Psychology? In *Semantics from Different Points of View*, ed. R. Bauerle, U. Egli, and E. von Stechow, pp. 1-14, Springer-Verlag, Berlin.

Pereira, F. C. N., and S. M. Shieber. 1987. *Prolog and Natural-Language Analysis*, CSLI Lecture Notes Number 10, Center for the Study of Language and Information, Stanford University, Stanford, California.

Perry, J. 1983. Situations in Actions, unpublished ms. of a lecture presented at the annual meeting of the Pacific Division of the American Philosophical Association, March 1983.

Perry, J. 1977. Frege on Demonstratives. *The Philosophical Review* 86(4).

Perry, J. 1979. The Problem of the Essential Indexical. In *Nous 13*, Indiana University.

Przymusinski, T. 1991. Three-Valued Nonmonotonic Formalisms and Semantics of Logic Programs. *Artificial Intelligence* 49, Nos. 1–3. 309–343.

Putnam, H. 1975. The Meaning of Meaning. In *Minnesota Studies in the Philosophy of Science: Language, Mind, and Knowledge*, ed. K. Gunderson, 7: 131–193. Minneapolis: University of Minnesota Press.

Putnam, H. 1977. Meaning and Reference. In *Naming, Necessity, and Natural Kinds*, ed. S. P. Schwartz, 118–132. Ithaca: Cornell University Press.

Quine, W. V. O. 1960. *Word and Object*, Cambridge, Massachusetts: MIT Press.

Quine, W. V. O. 1971a. The Inscrutability of Reference. In *Semantics*, ed. D. D. Steinberg and L. A. Jakobovits. London: Cambridge University Press.

Quine, W. V. O. 1971b. Quantifiers and Propositional Attitudes. In *Reference and Modality*, ed. L. Linsky, 101–111. London: Oxford University Press.

Quine, W. V. O. 1972. Methodological Reflections on Current Linguistic Theory. In *Semantics of Natural Language*, ed. D. Davidson and G. Harman, 442–454. Dordrecht: D. Reidel Publishing Co.

Reichenbach, H. 1947. *Elements of Symbolic Logic*. New York: Macmillian Co.

Reiter, R. 1980. A Logic for Default Reasoning. *Artificial Intelligence* 13(1–2): 81–113.

Rescher, N. and A. Urquhart. 1971. *Temporal Logic*. Vienna: Springer-Verlag.

Robinson, J. A. 1965. A Machine-Oriented Logic Based on the Resolution Principle. *Journal of the Association for Computing Machinery* 12(1): 23–41.

Rumelhart, D. E. and J. L. McClelland and the PDP Research Group. 1986. *Parallel Distributed Processing: Explorations in the Microstructure of Cognition, Volume 1: Foundations*. Cambridge, Massachusetts: The MIT Press/Bradford Books.

Russell, B. 1903. *The Principles of Mathematics*. New York: W. W. Norton and Company, Inc.

Russell, B. 1949. On Denoting. In *Readings in Philosophical Analysis*, ed. H. Feigl and W. Sellars, 103–115. New York: Appleton-Century-Crofts, Inc.

Ryle, G. 1949. *The Concept of Mind*. New York: Barnes and Noble, Inc.

Scott, D. 1970. Advice on Modal Logic. In *Philosophical Problems in Logic: Some Recent Developments*, ed. K. Lambert. Dordrecht: D. Reidel Publishing Company.

Shieber, S. M. 1986. *An Introduction to Unification-Based Approaches to Grammar*, CSLI Lecture Notes Number 4, Center for the Study of Language and Information, Stanford University, Stanford, California.

Shvarts, G. 1990. Autoepistemic Modal Logics. In *Theoretical Aspects of Reasoning About Knowledge: Proceedings of the Third Conference*, 97–109. Pacific Grove, California.

Skinner, B. F. 1984. Behaviorism at Fifty. *The Behavioral and Brain Sciences* 7(4): 615–621.

Soames, S. 1987. "Direct Reference, Propositional Attitudes, and Semantic Content," *Philosophical Topics*, 15(4): 47–87.

Stalnaker, R. C. 1976. Propositions. In *Issues in the Philosophy of Language*, ed. A. F. Mackay and D. D. Merrill, 79–91, New Haven: Yale University Press.

Stalnaker, R. 1984. *Inquiry*. Cambridge, Massachusetts: The MIT Press/ Bradford Books.

Stalnaker, R. 1993. A Note on Non-Monotonic Modal Logic. *Artificial Intelligence* 64(2): 183–196.

Stoy, J. E. 1977. *Denotational Semantics: The Scott-Strachey Approach to Programming Language Theory*. Cambridge, Massachusetts: The MIT Press.

Thomason, R. H. 1980. A Model Theory for Propositional Attitudes. *Linguistics and Philosophy* 4(1): 47–70.

Waltz, D. L. 1975. Understanding Line Drawings of Scenes with Shadows. In *The Psychology of Computer Vision*, ed. Winston, 19–91. New York: McGraw-Hill Book Company.

Warren, D. H. D., L. M. Pereira, and F. C. N. Pereira. 1977. PROLOG—The Language and Its Implementation Compared with LISP. In *Proc. Symposium on Artificial Intelligence and Programming Languages (ACM)*; *SIGPLAN Notices*, 12(8); and *SIGART Newsletter*, (64): 109–115.

Wittgenstein, L. 1953. *Philosophical Investigations*. Oxford: Blackwell.

Woods, W. A. 1975. What's in a Link: Foundations for Semantic Networks. In *Representation and Understanding*, ed. D. G. Bobrow and A. Collins, 35–82. New York: Academic Press, Inc.

Woods, W. A. 1981. Procedural Semantics as a Theory of Meaning In A.

Joshi, B. L. Webber and I. Sag (eds.), *Elements of Discourse Understanding*, pp. 300–302, Cambridge University Press, Cambridge, England.

Index